Climate Change and Arctic Security

Lassi Heininen · Heather Exner-Pirot
Editors

Climate Change and Arctic Security

Searching for a Paradigm Shift

palgrave
macmillan

Editors
Lassi Heininen
INAR at University of Helsinki
Helsinki, Finland

Heather Exner-Pirot
Observatoire de la politique et la sécurité de l'Arctique at CIRRICQ
Quebec, Canada

ISBN 978-3-030-20229-3 ISBN 978-3-030-20230-9 (eBook)
https://doi.org/10.1007/978-3-030-20230-9

This Palgrave Pivot imprint is published by the registered company Springer Nature Switzerland AG
The registered company address is: Gewerbestrasse 11, 6330 Cham, Switzerland

Contents

List of Contributors

Heather Exner-Pirot Observatoire de la politique et la sécurité de l'Arctique at CIRRICQ, Quebec, Canada

Wilfrid Greaves University of Victoria, Victoria, Canada

Lassi Heininen INAR at University of Helsinki, Helsinki, Finland

Salla Kalliojärvi University of Lapland, Rovaniemi, Finland

Sanna Kopra Arctic Centre, University of Lapland, Rovaniemi, Finland

Teemu Palosaari Tampere, Finland

List of Tables

CHAPTER 1

Introduction: Theorizing and Broadening Arctic Security—Towards the Environment and Climate

Lassi Heininen and Heather Exner-Pirot

Abstract This book assesses the construction of security in the context of climate change, with a focus on the Arctic region. It examines and discusses changes in the security premises of the Arctic states, from traditional security to environmental and human security. In particular, the book explores how environmental degradation and climate change impact, and have impacted and changed, security discourses and premises as well as theoretically discusses the possibility for another change: from circumpolar stability into peaceful change. Chapters cover topics including the ethics of climate change in the Arctic; the discursive transformation of the definition of security; the intersection between urban, climate and Arctic studies; China's emerging influence on Arctic climate security; to militarization and changes in the security nexus of the Arctic.

L. Heininen (✉)
INAR at University of Helsinki, Helsinki, Finland
e-mail: lassi.heininen@helsinki.fi

H. Exner-Pirot
Observatoire de la politique et la sécurité
de l'Arctique at CIRRICQ, Quebec, Canada

L. Heininen and H. Exner-Pirot (eds.), *Climate Change and Arctic Security*, https://doi.org/10.1007/978-3-030-20230-9_1

The book concludes with the question of whether a paradigm shift in our understanding of traditional security is possible, and indeed whether it is already occurring in the Arctic.

Keywords Security · Climate change · Environment · Arctic · Security paradigm

This book assesses the construction of security in the context of climate change, with a focus on the Arctic region. The theme of the book is inspired by an unresolved situation within modern society: never before has there been such consciousness of the importance of global ecology, and acceptance of climate change as a global, and man-made, security threat, requiring a human response through the traditional mechanism of the unified-state system which is maintained and supported by the current high degree of militarization. At the same time, there is a growing need for a paradigm shift in our approach to traditional (nationally-defined) security due to the combination of environmental pollution and rapid climate changes as a super 'wicked' problem.

All this makes us wonder if there are lessons learned from the environmental 'awakening'—the powerful movement which began in the 1960s and politicized the environment and pushed governments to take concerns of environmental degradation seriously. Or, from the public awareness and resistance movements in the 1980s, during the last peak of increased tension of the Cold War, against nuclear weapons and the threat imposed by the strategy of nuclear deterrence, channeled by the 'nuclear winter' scenarios anticipating a cooling of the Earth after a possible nuclear war (see Heininen in this volume). Would it be possible to establish the same kind of resistance and movements against contemporary 'wicked' problems (problems which are unsolvable due to their complex and inherently contradictory nature), in particular global warming, and also take into consideration strategic industries and energy users, such as the military as sources of pollution?

Among the fundamental questions raised in this volume are: Do individuals and states have common security interests, when it comes to the environment? If so, does environmental degradation and climate change present such common interests? Even if states take into consideration climate change as a security dimension, as it has been stated at the UN's Security Council (see Kalliojarvi in this volume), have they reconsidered their security priorities due to climate change? If so, do they count 'military pollution' when trying to recognize different sources of pollution/pollutants? Furthermore, is international environmental risk management possible, and if so, how will it address militaries/militarization, in particular the current high degree of militarization?

Or, is there a faith in technology, such as geoengineering/climate engineering, that technology will come and save us? Could this be the reason for a delay in action by unified-states? If so, then it would be fair and ethically right, in order to have legitimacy, to have an open discussion on geoengineering as a proper and accepted method to (try to) solve the problem.

Or, is it that in light of environmental degradation, as well as in that of climate change, people must consider their own 'security priorities', and not accept the use of traditional military ways. Behind this is the well-known Ulrich Beck's theory on the risk society that modern societies consciously take bigger risks concerning the environment which is clearly shown by the increasing use of nuclear power despite lethal accidents, as well as environmental degradation by the military.

Changing the Paradigm

In security studies, there are discourses, premises and paradigms of security, and changes in these premises and paradigms, as well as discussion on who the subjects of security are. In recent decades we have seen new discourses, such as on environmental and human security, and emerging shifts in premises; but not (yet) a shift in paradigms.

When it comes to the post-Cold War Arctic, there are special features of security, such as militarization, nuclear weapon structures, and significant changes in security premises from military to environmental security, parallel to the geopolitical shift from confrontation (of the Cold War) into cooperation (of the post-Cold War). This was much due to long-range pollution and nuclear safety, and the urgent need for functional cooperation on environmental protection, as is discussed in the book.

Although there has been a shift in security premises, it has not resulted in either nuclear disarmament nor a paradigm shift, yet. Due to the security dimension of pollution and climate change, as grand environmental challenges and/or wicked problems, there are on the one hand, new and more subjects of security which require a further and deeper change in problem definition on security premises, and on the other hand, consequently, a growing need for a constant shift in security paradigm(s). Correspondingly, this paves the way to argue that there is a need, even room, for a more permanent change in the problem definition of security paradigm(s), including disarmament, demilitarization and how threat/enemy pictures are constructed.

The focus of this book is climate change and global security, as cross-cutting issues dealing with environmental and human security discourses, in general and in particular in the Arctic context. The security nexus of the post-Cold War Arctic, transferring from military confrontation to cooperation to environmental security includes special features of security, such as nuclear safety (due to pollution) causing a change in security premises, and rapid climate change requiring a paradigm shift. The post-Cold War Arctic faces the 'Arctic paradox', grand environmental challenges and wicked problems (the combination of pollution and climate change).

We ask in our publication, whether there is a need and interest to change the problem definition of security premises and paradigms? We interpret that rapid climate change and the *Anthropocene* are global factors, and ask whether they can promote a peaceful change, though it is not assured, and whether it would be logical to consider that? If so, that discourse includes an option of a paradigm shift. Finally, we search for arguments for, and indicators of, a paradigm shift when it comes to security in the environmental and Arctic context.

Summary of Chapters

Our questions are addressed in six chapters, in addition to the short introduction and conclusion, which each look at different aspects of climate change and global and/or Arctic security, and how they are evolving and intertwined.

Salla Kalliojärvi provides a theoretical underpinning for our discussion in her chapter entitled *Age of Changes: Threats of Climate Change and Its Meaning for Security.* As she explains, the security implications of climate change have attracted increasing attention of academics, policymakers and security practitioners since the 2000s. Climate change is often perceived as a 'threat multiplier' that has the potential to amplify already existing security threats from poverty to large scale displacement of people to conflicts, pushing the responsive capacities of many governments and societies to their limits. A great number of states, among with many international institutions, have identified climate change as a national and international security threat. The problematization of climate change as a security issue has not followed unitary logic, but is varying greatly in the assumptions of who or what should be secured, from what threats and by what means. The different ways of problematizing climate change as a security issue entail different sets of policy recommendations and construct different understandings of security.

Kalliojärvi presents a poststructuralist and post-Marxist analysis of the climate security discourse, investigating what kind of security climate change is said to endanger and how the security is provided. Her analysis focuses on the discursive construction of/and transformations in the meaning of security, and argues that the prevailing understanding of security is highly centralized around the secondary implications of climate change that emphasizes the need of adaptation measures but misses acknowledging the importance of emission reductions in maintaining security. Due to its unique exposure to climate change, the Arctic is highly dependent on the security policies adopted on a global scale. Kalliojärvi argues that the current climate security discourse is more likely to result in adaptive strategies than effective mitigation efforts that are preparing for an iceless future rather than aiming to tackle the global warming.

Sanna Kopra examines the role of China in our evolving paradigms of international security, especially with regards to the Arctic, in her chapter on *China, Great Power Responsibility and Arctic Security*. As China increasingly identifies itself as a great power in world politics, "China threat theories" and speculations as to whether its rising status will change discourses, premises and paradigms of international security, have grown. When it comes to the Arctic, China's growing interest in regional affairs has raised questions of its potential risks to regional peace and security. As China is now the largest carbon emitter in the world, it presents a tremendous challenge to climate change mitigation and human security regionally and globally. Based on a premise that great powers have a unique responsibility to uphold international peace and security, Kopra examines China's emerging great power status and its implications for Arctic security in general and in the context of international climate politics in particular.

Teemu Palosaari, in his chapter on *Climate Change Ethics in the Arctic*, looks at how the key questions of global climate change ethics—moral responsibility, distribution of benefits and burdens, and environmental justice—appear in the Arctic discourse. As Palosaari describes, a significant share of the world's as-yet unexploited oil and gas resources are at the bottom of the Arctic Ocean. As the sea ice is melting, coastal states and energy companies are viewing these resources with interest. Using them would, however, accelerate climate change, leading to a heated debate on whether the new Arctic oil and gas reserves should be utilized or left untouched. Palosaari's main argument is that it is becoming increasingly difficult for governments, businesses and decision-makers to ignore climate ethics in the Arctic. The way that problems and solutions regarding Arctic oil and gas are defined and promoted has global significance. Arctic glacier and sea-ice melt have both symbolic value in global climate politics as well as a direct impact on the rise in sea level around the world.

Will Greaves takes a rare and overdue look at the intersection between urban, climate and Arctic studies in his chapter on *Cities and Human Security in a Warming Arctic*. As he points out, humanity has became an urbanized species, with a majority of people around the world living in urban areas. In the same period, the scale of humanity's impacts upon the natural environment, most notably through human-caused climate change, confronts people and societies around the world with fundamental challenges to their continued ability to live where and how they wish. The nexus of urbanization and climate change poses fundamental questions relating to how cities should be understood in relation to the protection of people from environmental insecurities.

Greaves develops an argument that cities and smaller urban centres are paradoxical for questions of climate and environmental security. Urbanization represents an important development for improving global environmental sustainability by reducing individuals' carbon footprints and resource use-intensity, but are also significant consumers of resources and producers of pollution and waste that contribute to conditions of global ecological crisis. Moreover, cities simultaneously offer greater opportunities for individuals to meet their basic survival needs and can provide some resilience for people in the face of natural disasters. However, many cities are also vulnerable to disruptions caused by climate change, and face growing challenges as a result of changing ecological conditions. Thus, cities are paradoxically implicated in both securing people against climate change while making them more reliant upon industrial processes and systems that directly contribute to climate change. This argument is developed through examples from urban communities in the circumpolar Arctic region, which is not only a highly urbanized space, but also the global region experiencing the fastest and most acute impacts of climate change.

Heather Exner-Pirot explores more traditional concepts of security as applied to the Arctic in her chapter *Between Militarization and Disarmament: Challenges for Arctic Security in the Twenty-First Century*. The period of regionalization that the Arctic has enjoyed since Gorbachev's famous Murmansk Speech in 1987 has brought stability and peace. Ironically, however, it has not led to disarmament, despite the initial premise of developing the Arctic region into a 'Zone of Peace', with an Arctic Council focused on demilitarization and arms control issues. Exner-Pirot asks why issues of militarization have fallen so far off the Arctic states' political agenda.

Her chapter provides an historical overview of disarmament and demilitarization discussions in the Arctic, including coordination around cleaning up nuclear and radioactive waste. It then turns to more contemporary issues around militarization: the threat of an arms race in the region, an absence of fora to discuss issues of military security; and the geopolitical dynamics, especially concerning Russia, that make Arctic demilitarization efforts so challenging. Exner-Pirot concludes with an assessment of the short and long-term security trends in the region and strategies for more explicit peace-building, including demilitarization.

Finally, **Lassi Heininen** provides an overview of the environmental and peace movements in the Arctic and their intersection, in his chapter *Before Climate Change, 'Nuclear Safety' Was There: A Retrospective Study and Lessons Learned of Changing Security Premises in the Arctic*. The arms race between the two major nuclear weapon powers in the Cold War was so hectic that there was neither time nor expertise, nor perhaps not even reason, to think about cleaning-up after the 'party', i.e. eliminating the environmental damage that resulted from the use of large territories and sea areas for the military, constant testing and training, nuclear (submarine) accidents and leakages, experimental deployment and testing of weapons. Thousands of tons of garbage, toxics, (solid and liquid) radioactive waste, and pollution were left behind. Due to growing concern for the environment and the environmental awakening in the 1980s and 1990s, the 'militarized' Arctic started to become 'environmentalized' and security was 'reconceptualized'—the environment started to matter. Still in the 2010s, however, these problems persist.

Based on the transformation towards environmental and societal security Heininen first, (re)conceptualizes significant changes in Arctic geopolitics and Arctic security, as well as combines geopolitics and security with each other; second, as outcomes he describes and analyzes the current phase of the Arctic security nexus, and that of Arctic geopolitics, with the environment and climate, as well as comprehensive security, much recognized as drivers, though in cases of the Arctic littoral states with the emphasis of state sovereignty and its defense in the Arctic.

***

This book project continues the Palgrave Pivot peer-reviewed publication series (based on the Security of the Arctic panel) by the University of the Arctic and Northern Research Forum joint Thematic Network on Geopolitics and Security. The two previous volumes include *Future Security of the Global Arctic: State Policy, Economic Security and Climate* (Palgrave Macmillan, Palgrave Pivot, 2016), and *Security and Sovereignty in the North Atlantic* (Palgrave Macmillan, Palgrave Pivot, 2014), both edited by Lassi Heininen.

CHAPTER 2

Age of Changes: Threat of Climate Change and Its Meaning for Security

Salla Kalliojärvi

Abstract Climate change has been increasingly defined as a threat to the security of a wide range of referent objects from individual to the planet and humanity as a whole. Climate change is often seen as the macro-driver of a wide range of other threats that are accelerated by the extent of which the climate continues to change. By adding pressure on the institutional capacities of states, decreasing the quality, quantity and access to vital natural resources, and by triggering large scale migratory movements, climate change is assumed to increasingly foment instability and conflicts, which in turn are seen to deteriorate global sustainable development that forms the very heart of mitigation and adaptation efforts. Regardless of this growing and often alarmist security framing of climate change, ambitious and effective climate policies are still lacking. Traditionally security is conceived as a matter of high politics that is associated with the core values and survival of the state and the legitimization of emergency measures. Military force or balance of power politics and containment have long been seen as the most crucial ways to control insecurity. The threat of climate change is characterized with a new kind of uncertainty and complexity that cannot be

S. Kalliojärvi (✉)
University of Lapland, Rovaniemi, Finland

L. Heininen and H. Exner-Pirot (eds.), *Climate Change and Arctic Security*, https://doi.org/10.1007/978-3-030-20230-9_2

successfully secured against through military power and containment. Building on poststructuralist theories, this chapter theoretically analyses how the meaning of security is constructed through hegemonic struggle, and how the interpretations of climate change as a threat or a multiplier of threats affect the understanding of security.

Keywords Climate change · Security politics · Environmental security · Critical security studies · Poststructuralism · The United Nations Security Council

Introduction

During the last decade the attention paid to the security implications of climate change has increased significantly. Even though the potential and threatening implications of changing climate were discussed already in the early 1970s (Stripple 2012, p. 185), it was only at the beginning of the twenty first century, that the labelling of climate change as a security threat started to proliferate (Rothe 2016; Oels 2012; Trombetta 2008; Dalby 2013). Unprecedented public attention was given to the matter in 2007 when the Nobel Peace Prize was awarded to the Intergovernmental Panel on Climate Change (IPCC) and the former US vice president Al Gore for their efforts in obtaining and disseminating information on climate change and as for the first-time, climate change was brought to the discussion of the United Nations Security Council (UNSC). Since then, the UNSC, along with many other key international institutions, have discussed the regional, international and global security implications of climate change on several occasions, and a great number of states have recognized the threat of climate change in their foreign and national security planning (McDonald 2018; Scott 2015). Regardless of the growing and often alarmist security framing of climate change, there has not yet been any coordinated and ambitious strategies in addressing the issue, but only a widening of the gulf between rhetoric and action.

In the academic literature the relationship between climate change and security has been subjected to a variety of discursive lenses such as national security, human security, environmental security, conflict and risk, that all problematize different aspects of the relationship, rely on different premises and produce different kind of discursive effects. The discussion about the relationship has been characterized with a deep division between those arguing that the logic of security is fixed and inflexible, and the others suggesting that climate change necessitates a transformation of the security field towards a more precautionary, global and integrated approach. This division has its roots in the debate that during the last few decades has taken place within the field of Security Studies over the normative and analytical appropriateness of broadening and deepening the security agenda to include non-state actors and non-military issues, such as the environment. Where some scholars have opposed the broadening of security agenda or feared that the term 'security' evokes a state and military centered set of confrontational practices and attitudes that hider international cooperation, and risks leading rather to the militarization of environment than greening of security (Deudney 1990; Detraz and Betsill 2009, p. 314; Käkönen 1994), others have identified opposite transformations in the values and practices traditionally associated with security (Rothe 2016; Trombetta 2008; Oels 2012). For the latter group, rather than the security language solely transforming the method dealing with climate change, it is the discourse of climate change that has impacted on the definition and governance of security. This chapter will follow this line of thought by examining how the interpretation(s) of climate change as a threat are producing and contesting the meaning of security. In doing so, this chapter will introduce a way of exploring the ways in which climate change is impacting on the meaning(s) of security by drawing on the insights of a poststructuralist discourse theory and Ernesto Laclau and Chantal Mouffe's revision of Gramsci's concept of hegemony. Diverging from traditional approaches to security that defines it as an object of the world, security is understood as a socio-political construction that reflects the radical contingency of meaning. As Karin Fierke (2015, p. 27) points out, in a world that is undergoing dramatic change, theory cannot mirror a static reality but must be a source of reflection and critique. The chapter is based on analysis of the UNSC's debates on climate change, which serves as an empirical illustration of the discursive struggle climate change is provoking over the meaning of security.

The UNSC is strongly reflective of the dominant idea of security that prevailed in the aftermath of the two World Wars, comprising the core ideas of realism and liberalism (see e.g. Fidler 1996, p. 415; Weiss 2011, pp. 21, 118). From a Discourse theoretical point of view, the Council can be seen as a part of the hegemonic discourse on security. Regardless of the numerous attempts to reform the Security Council, there has not been any radical structural changes in the Council since its establishment in 1945. The predominant focus of the Council has been on conflict management and hard security interventions, but in recent years the Council has shown intention to move beyond a reactive approach and broadened the scope of identified threats to include issues such as HIV/AIDS and climate change. Climate change, however, has not been unanimously accepted into the agenda of the Security Council. The Council is highly divided in the view of the legitimacy of the Council to debate and act on the issue of climate change. Even the participants welcoming the discussion differ greatly over their understanding of the nature of the threat posed by climate change or what constitutes an appropriate way of responding to it. The Security Council thus offers an interesting platform to study the understanding(s) of security in international politics and serves as a reflection of the predominant views prevailing in the international field. It, of course, is far from being exhaustive or all-encompassing, but nevertheless a relevant one in seeking an understanding to the question of: 'how does climate change impact on global security politics?'

Security as a Discourse: The Ontology of Security

Ever since of the establishment of the nation-state system in the Treaty of Westphalia in 1648, security has been principally understood in relation to state sovereignty and territorial integrity that are always threatened by the possibility of war (see Baylis 2014, p. 496). In this view security is ultimately defined by state survival, as the sovereign state—conceived as the principal actor in the international system—is seen to be acting under anarchy that forces states to seek security by maximizing their material power relative to other states (Mearsheimer 1990, p. 12). Military force is thus understood as forming, not the only, but the most crucial part of the security structure, as the ultimate mechanism to maintain security is, paradoxically, that of resort to war. While war and the threat of the use of force still remain as main drivers of security policies of many states and institutions, especially after the end of the Cold War, new threats such as civil conflict, terrorism, drug trafficking, international organized crime, resource scarcity, environmental degradation, illicit or uncontrolled migration, poverty, AIDS and now climate change, have emerged to the security agenda. These new threats are characterized by a new kind of interconnectedness and complexity that make them less predictable and impossible to be tackled through containment and separate security practices. In the academic literature the changing scope of security is often identified to be, simultaneously, a byproduct of globalization and a threat to it (Dalby 2015, p. 431). In the Arctic context, this dualism of globalization (Heininen 2016, p. 20) is often referred to as the 'Arctic paradox' (Palosaari 2012; Palosaari and Tynkkynen 2015, p. 98).

The broadening of the security agenda, along with the end of the Cold War confrontation, brought up many fundamental questions about the meaning and focus of security policy. Questions such as what makes a particular problem a security issue, should the state still be the referent object of security policies, are the traditional security practices and instruments still relevant, or how are the security relations transformed, have become concerns of many academics and politicians during the last few decades. By rejecting the core assumptions of positivism that forms the backbone of traditional security studies, and analyzing the processes by which security and insecurity are produced, and the political ramifications of this process, many scholars have been able to point out how the relationship between who protects who, from what and how, is fundamentally a political one (Fierke 2015, p. 32; Campbell 1998; Der Derian and Shapiro 1989, p. 12; Huysmans et al. 2006; Crawford and Lipschutz 1997; Dillon 1996). This aspiration to excavate the logics of security at work in world politics has not only shifted the focus of security analysis from the investigation of empirical facts towards the analysis of its conditions of possibility, but has also paved the way to explore the processes of contestation and change. This chapter follows this path and introduces a way to explore how the discussion of the security implications of climate change impacts on security politics. By building on Ernesto Laclau and Chantal Mouffe's poststructuralist and post-Marxist Discourse Theory[1] that has its root in structural linguistics, French poststructuralism, and psychoanalysis, the chapter builds on a thought that the transmission of meaning through discourse is the driving force behind any social change. Central here, is the understanding of discourse as an ontological and epistemological category, which means that all social phenomena are mediated through discourse, and no meaning pre-exist this relational complex (Laclau 2005, p. 68; Laclau and Mouffe 2001, pp. 94–95). This, of course, does not mean that there is no world external to thought, but that all meanings are discursively constructed, and all power rests on meaning and legitimacy.

The Discursive Struggle

Discourse is understood as a relational complex of signifying sequences where signs and material objects, language and pragmatic aspects of action have been temporarily woven together, and produced and reproduced through a pattern of repeated articulation of privileged signs that partially fix the meanings (Laclau and Mouffe 2001, pp. 91–100; Rothe 2016, pp. 50–51). Owing to the fundamental instability of language, the fixation of meaning can never be permanent but discourses are always engaged in a constant struggle over the 'right' representation of the ways of being in, and acting towards, the world (Laclau and Mouffe 2001; Jørgensen and Phillips 2002, pp. 26–27). Discourse, therefore, is "constructed in and through hegemonic struggle that aim to establish a political and moral-intellectual leadership through the articulation of meaning and identity" (Torfing 2005, p. 15). As such, security discourses always both depend on and sustain particular representations of the world. How the state leaders define security relies strongly on how they conceptualize the state and its place in the world, and inherently, the desirable nature of the political community (Nabers 2015, p. 163; McDonald 2002, p. 285). As pointed out by Lene Hansen (2006, p. 35) security discourses are "characterized by a dual political dynamic: they invest those enacting security policies with the legitimate power to undertake decisive and otherwise exceptional actions, but they also construct those actors with a particular responsibility for doing so."

From a Discourse Theoretical point of view, the institution of a social identity is always an act of power of creating relations of "equivalence" between subject positions and disparate demands (Laclau and Mouffe 2001, pp. 85–88, 128; Laclau 2005, pp. 72–77). The chain of equivalence unites different societal demands under a hegemonic demand—i.e. an empty signifier—that represents them all. In this way, empty signifier represents the attempt to "fill" the unavoidable lack of full closure of the social (Laclau 1996, p. 53). As I will elucidate later, sustainable development represents such a privileged sign in the evolving hegemonic discourse of security, resulting from the dislocative effect of climate change. Sustainable Development signifies a logically unattainable universality assuming a totalizing function, thereby making community-building possible by symbolizing the equivalential chain as such.

Curial in the creation of equivalence of demands is the naming of 'others' who are presented as blocking the identity and attainment of the demands. The construction of common enemy downplays the particular differences between identities and demands brought together by the chain of equivalence, and modifies them to reflect their more universal character (Griggs and Howarth 2014). In other words, equivalence makes differences equivalential with regard to something identical underlying them all, the excluded one, a commonly antagonized 'Other' (Laclau 2005, p. 78). When a particular demand manages to assume the representation of a common or universal interest of a political community, it becomes hegemonic. Universality, hence, can only exist in the form of hegemony where particularity is transformed into the representation of universality (Laclau and Mouffe 2001, pp. 119, 125–127; Laclau 2014, p. 6). Once discourses have become hegemonic the social practices they structure become naturalized or sedimented and receive a taken-for-granted status. In a contrast to this, the logic of difference attempts to break the chains of equivalence by enhancing the differences and trying to relegate the antagonistic divisions into the margins of society (Howarth 2000, p. 107; Laclau and Mouffe 2001, pp. 113–120). This (re)particularization separates the subjects from their equivalential chain weakening the dichotomization of the discursive space.

As exemplified above, understanding of discourse as the primary terrain of the constitution of any objectivity, means that no danger is an objective condition that exist independently of those who it is threatening, but the identification of a threat always comes with identifying those who are rendered secure. This point is well exemplified by David Campbell (1998) in his study of the United States' foreign policy, where he convincingly shows how the interpretations of danger produces and reproduces the identity of the United States by establishing the boundaries of inside/outside, domestic/foreign, civilized/barbaric. Identifying or 'naming' something as a threat is, therefore, not just providing a label to a pre-existing object, but a process of identity construction of various subjects and their positioning in relation to each other (Shepherd and Weldes 2008, p. 234; see also Butler 1992, p. 2). The shift in understanding security as a socio-political construction, in this way, overcomes the tendency of traditional security studies to take security as a property of independently existing objects of the world trapped in a 'security dilemma', by replacing the universal essence of security with discursively formed meanings that are more or less institutionalized. The security environment is not a pre-given or predetermined, but the ascriptions of danger have performative power that inscribes social and political relations. Hegemony, therefore, is much more than just the dominance of one state over others, as often understood in the field of International Relations.

In this light, analyzing the rise of climate change as a threat is less a matter of investigating the broadening and deepening of security and more of rethinking its rationale. Understanding the impact of climate change on security politics means studying the transmission and transformation of meaning through discourse in order to reveal how particular discursive forms of power naturalize certain processes of institutionalization and dismiss others. This means exploring and revealing the wider order of meaning that security is embedded within and identifying the points of discursive struggle. The security environment today is conceived to be more complex and interdependent, composing a mixture of intertwined national, international, local, regional, domestic, and global security problems that breaks the tradition of seeing security mainly as outwards-oriented (see Cavelty and Balzacq 2017, pp. 29–31). Security is no longer primarily about geographically identifiable threats and battles against known enemies, but increasingly characterized by a borderless society-based narrative of vulnerability and the stabilization of technical and societal systems designated as vital to collective life. Instead of being interpreted as a 'single problem' that unites humanity into one political community, climate change is often presented as a condition that, in the name of stability and security, forces differential transformations in societies depending on the geographical and socio-economic positioning.

The next sections will show how security against the effects of climate is understood not only as a freedom from dangers, but as a capability to adopt. Undisrupted economic growth is taken both as a necessity for security and as a seed of growing insecurity which constructs a demand for sustainable development. Sustainable development is presented as an antidote against vulnerabilities, which if let unmanaged can turn into security threats. Along with states and international institutions, private sector and civil society are seen as crucial security actors. Global governance is not a danger but necessity for maintaining sovereignty threatened by the forces outside of the governing structure, such as unsustainability and criminal groups.

Climate Change, Security and the United Nations Security Council

The concept paper prepared by the United Kingdom and circulated around the member states before the first ever meeting of the UNSC on the security implications of climate change, identified climate change as an accelerator of vulnerability that increases the risk of social instability and conflict, in and between states (UN 2007). The concept paper justified the discussion by the Council's commonly agreed commitment[2] to conflict prevention and peace-building, and called for promotion of sustainable development in order to address the root causes of conflicts. In the meeting, the causal link established between aggravated climate impacts and increased conflict risks was strongly opposed by some of the member states, such as China, Russia and many of the Group of 77 members, who rather saw the attempt of integrating climate change into the agenda of the Security Council as an act of aggression and concern of security. Meanwhile many of the small island states stressed that climate change for them poses a far greater threat than any conflict or war. After being disappointed by the nature and outcome of the debate, Pacific Island States took the issue to the General Assembly, where a resolution 63/281 expressing a deep concern for the possible security implications of climate change was adopted in 2009.

In 2011, in recognition of the growing security concerns posed by climate change, the Security Council held a second open debate on the impact of climate change on the maintenance of international peace and security, under the German presidency. In comparison to the 2007 meeting, the number of both the participating states, and those in favor of an active role of the Security Council in addressing climate change-related insecurity, increased notably (Cousins 2013, p. 203). Even as the dispute over the mandate and role of the Security Council was still central to the debate, the rejection of the link between climate change and security that was still evident in 2007 was nearly absent in 2011. While the Council was still unable to produce a resolution, it managed to pass a modest Presidential Statement (UN 2011b), where it recognized that "possible adverse effects of climate change may, in the long run, aggravate certain existing threats to international peace and security" and requested the Secretary-General to report contextual information on possible security implications of climate change. The statement also acknowledged the threat of loss of territory for small-island states due to sea-level rise, but missed making any specific recommendations for action. In 2015, sea-level rise and the special security challenges facing small island states were brought up on the Security Council's table again by New Zealand. The debate reinforced the view of climate change as a multiplier and source of many global threats and created political space for articulating climate change and the inaction as grave threats themselves to the national security of many small island states and to the future of the whole humankind.

Whereas for states anchoring the meaning of security to the state centric powerplay determined by anarchy, the integration of climate change into the agenda of the Security Council risks losing sovereignty and the right to pursue economic development, for many small island states the integration is a way to preserve the very same things, as it is not the encroachment of the Security Council into areas of sovereign state authority but the physical encroachment of the rising seas and the loss of livelihoods resulting from climatic changes that constructs the most pressing concern. For them, sovereignty is not threatened by foreign intervention but by the lack of global cooperation. Without global cooperation, the interconnectedness, complexity and globality of the security implications of climate change is seen to reduce the ability of the states to provide security. As Mr. Beck (UN 2011a), the representative of Palau pointed out at the UNSC's meeting in 2011: "While the causes of this threat are novel, the effects, which endanger the sovereignty and territorial integrity of Member States, fit squarely within the Council's traditional mandate." Instead of drawing on the interventionist discourse, the enhancement of security in the frames of climate change is best seen to be pursued through partnership between multiple actors form different levels and sectors of societies. The multidimensional risks posed by climate change are requiring multidimensional responses that necessitates the contributions of the private sector and civil society partners. Largely associated with voluntarism, adjustability and empowerment, partnership does not mean eroding of states' authority or violating sovereignty, but is conceived to be strengthening them against the eroding effects of climate change.

The discursive shift toward the weakening of the respect for non-interference and the rise of what some scholars have called 'contingent sovereignty' especially in Western security politics is not, however, a consequence of the climate change discussion, but can be traced back to the end of the Cold War Confrontation (Duffield 2007; Elden 2006).

Instead of absolute sovereignty, state sovereignty became increasingly understood as entailing universal obligations or responsibilities that states have toward their own citizens and certain norms of behavior. The failure to meet these responsibilities started to form a legitimate reason to break the state's sovereign right to non-interference as the responsibility to protect the civilians in these situations devolved to the international community—most commonly associated with the UNSC (Bellamy 2010, pp. 362–363). In this discursive shift the traditional idea of security being best pursued "through a society of sovereign states that enjoy exclusive jurisdiction over a particular piece of territory and rights to non-interference and non-intervention" (ibid., p. 362) started to transform towards an idea of intervention as capacity-building, simultaneously, shifting security discursively from a concern of states with defense towards the interventionist discourse of global human rights protection (Chandler 2012, p. 218; 2015, p. 76; Piiparinen 2012). Interrelating with the discourses of Liberalism and neo-Liberalism, security became increasingly centered around the life of the population instead of simply the safety of the sovereign or the security of a territory.[3] This shift is well exemplified by the former UN Secretary General Ban Ki-Moon in his statement at the Security Council's meeting in 2015: "Governments earn, sustain and strengthen their sovereignty by being accountable to their people, upholding human rights, ensuring the rule of law and practicing inclusive governance. [...] Sovereignty remains a bedrock of international order. But in today's world, the less sovereignty is viewed as a wall or a shield, the better our prospects will be for protecting people and solving our shared problems" (UN 2015).

While constructing the most radical attempt to rearticulate the meaning of security, the articulation producing climate change as a 'direct threat' or a 'weapon' that constructs an existential threat to those identifying themselves as victims, is not shared by many within the Council. Instead, the indirect consequences of climate change such as increased human insecurity and statelessness bearing a witness of the dangerousness of climate change in the articulation of the victims, are defined by the majority as the threats that the international community needs to be addressing. Coalescing with the claim that development requires security and without security there cannot be development that has increasingly been generalized since the end of the Cold War (Duffield 2007, p. 1), the articulation of climate change as a threat multiplier establishes development as a practice of security. Climate change is seen not only to aggravate and foment conflicts, but to increase 'climate-change-related poverty' that breeds 'recruitment by rebel or terrorist groups', causes 'statelessness', 'migratory pressure' and 'cultural mortality' that are linked with lost coherence, hatred, instability and alienation that in a globally integrated and networked world increases the risk of spread of instability and violence.

As it will be elaborated in the next section, it is not simply the changes in the climate and weather but the societies' ability and ways of responding that is the primary concern. Climate change is transitioned from an environmental problem to a problem of governing and maintaining stability where helping the most vulnerable to develop, is taken out of a mere economic—and even normative—context and linked to the national interest of preventer states and the maintenance of global security.

Climate Change as a Threat Multiplier: Merging of Securities

Connected to the shifting of sovereignty discourse, the securitization of development after the terrorist attacks of September 11, 2001 within the Western nation-states, came with a growing attention to the socio-economic roots of threats such as terrorism, state fragility and mass migration, challenging the direct spatial and temporal dimensions of security (Duffield and Waddel 2006; Duffield 2006, 2007). The underlying storyline in the merging of development and security, as exemplified by the former UN Secretary General Kofi Annan, is that "[a] more secure world is only possible if poor countries are given a real chance to develop. Extreme poverty and infectious diseases threaten many people directly, but they also provide a fertile breeding ground for other threats, including civil conflicts" (cited in Stern and Öjendal 2010, p. 5). Behind this line of thought is a belief that the creation of liberal democratic societies with market economies will engender peace and stability, and secure the already developed societies from the ripple effects of underdevelopment. This discursive interplay between security and development represents an important genealogical root of the climate security discourse that designates climate change as a multiplier of already existing threats, a serious challenge to the development of many states and societies, and a primary threat to global sustainable development (see Methmann and Rothe 2013, p. 189; Gilmore 2017).

In this discourse, by negatively impacting on the living-conditions of individuals and the institutional capacities of states, climate change is seen to endanger the socio-economic foundations understood as necessary for stability, order and peace. In order to ensure functionality and security, the state and its populations need sufficient access to the critical resources underpinning national security such as water, food, transport, and energy systems. Climate change is seen to put additional pressure on these resources and potentially degrading the state's ability to deal effectively with the threats to human security, simultaneously increasing the risk of social upheaval, loss of legitimacy and violent conflict, that in turn, are linked to a broad range of destabilizing and threatening effects such as terrorism, organized crime and mass migration. In this regard, the failure of a state to provide the access and hence human security to its citizens that forms a threat to the state itself, also constructs the society as a future trouble spot or a risk group that threatens global stability and those socio-economically better equipped. The link established between the degree of vulnerability and risk shifts the focus on the inherent capabilities of vulnerable communities and to the institutional structures of states, where the multiple and varying sources of vulnerabilities are tied together through their common underminer of climate change. Human security issues have, thus, become a factor in assessing the stability of governments and of the state itself, as well as an essential part of the idea of global security.

Highly influenced by a neo-Malthusian narrative, poor, less developed societies and fragile and weak states are seen as the most potential breeding grounds for conflicts as it is the lack or fragility of state institutions that account for failed adaptation and provides conflict prone conditions. According to Betsy Hartmann (2013, pp. 154–156) the climate crisis narrative is very much like old wine in a new bottle, where the neo-Malthusian models of environmental conflict developed after the 1980s are remodeled into the climate change arena, where they continue to serve the counterinsurgency strategies of Western defence interests. The neo-Malthusian narrative locates the resource related conflicts primarily to the so-called developing states from where, if not rightly governed, they can spread to all over the world, and form a threat to the identity of developed states as a coherent and stabile (Hartmann 2010; Bettini 2013, p. 209). While mostly advocated by Western states, the articulation of climate change as a threat multiplier links climate change to the necessity of prevention that enforces the logic of security based on risk management and institution building. Governance through risk is a central part of defense strategy, as strengthening the adaptive capacities of these communities and states is constructed as a crucial tool in preventing the actualization of threats. Instead of a punitive and coercive measures, climate change as a threat multiplier stresses the importance of all level and sector wide partnerships and cooperation that are based on the rationales of empowerment and adjustability. Especially, the private sector is often identified as a crucial partner in providing security against the proliferating security impacts of climate change. The inclusion of the private sector into the security field, or what some scholars call as privatization of security, has not, however, resulted from climate change debates, but has its roots in the proliferation of non-state armed actors, including rebel movements, insurgents, guerrilla groups, and mercenary forces over the last two decades (Lilly 2000; Mandel 2002; Krahmann 2003).

Influenced by ecological thinking, security against the multiplying effect of climate change is understood to be keenly determined by complex interaction in the articulation of climate change as a threat multiplier. To avoid the unintended negative consequences of adaptation, the understanding of environmental, social and political contexts and their interaction is seen as essential, and legitimating the claim for adjusting the actions ensuring sustainable development based on particular local settings. In all levels—human, state, international and global—sustainable development is set as a precondition for a secure future that is threatened by unsustainable and non-developmental systems. Sustainability is presupposed to stop climate change from accelerating in a dangerous manner, whereas development is conceived as a prerequisite for achieving and building adaptive capacities, which in turn are a necessity in protecting the well-being and lives of humans and states, and thus maintaining security on a global level. Rather than simply meaning effectively balancing social, economic and environmental objectives in decision-making, sustainability is linked to the enhancement of human security that is vital for the maintenance of global stability and order (see Duffield 2006). In a similar vein, based on his studies, Mark Duffield (ibid., p. 19) has argued that sustainable development "defines the type of 'development' that is securitised in human security". Also, the EU Global Strategy exemplifies this point by defining sustainable development as "the condition for security, stability, social justice and democracy – abroad as much as at home" (Gavas et al. 2016, p. 2).

In previous sections I have tried to elucidate how it is not climate change but the lack of sustainable development that takes the form of antagonism, and how sustainable development comes to symbolize the equivalential chain of diverse demands. A crucial feature of the sustainable development signifier as privileged signifier is its ambiguity and vagueness that enables it to work as an answer to multiple different security demands. The next section will illustrate how the ambiguity and vagueness of the signifier enables it to be accepted as a universal tool that is holistic and precautionary, but leaves open the ways in which its implementation actually happens, leaving it compatible with many different security articulations and discourses.

The Traditional Heritage

As shown by Sanna Kopra (in this volume), great power states are usually conceived to bear greater responsibility in maintaining international peace and security. In both articulations put forward in the UNSC's meetings, the one identifying climate change as a direct threat and the other producing it as a threat multiplier, the subject position of a preventer is linked to a developed, responsible state that is expected to act in concert with other preventers in ensuring the security of the victims and maintaining international security. The environmental changes resulting from climate change are seen to engender stress to societies, which without requisite adaptation measures and preventive action on behalf of the preventers, will cause mass migration and violent behavior, as war and violent conflicts are seen as likely ways to secure the possession of scarce resources and livelihoods. While the articulation of climate change as a threat multiplier is shared by many who identify themselves as vulnerable, the legitimacy of the Security Council to address climate-related security threats outside of a conflict context is not accepted. For them the Security Council represents the international structure from the time of its founding, owning the old power structures and inheriting military and sanction-based tools, which are not appropriate in handling climate related non-traditional security threats. Especially the exclusive nature of the Council which concentrates power to the 'largest emitters' makes it unacceptable to the Council to take action toward global and complex threats of climate change. Under the power structure favoring the largest emitters, the Security Council is not trusted to act justly or effectively enough.

For many developing states, the Security Council is not seen to be representing the global community but the nationalistic interests of the permanent members. The increasing military expenditures of the permanent members is used to exemplify the hypocritical position of the Council in debating the non-traditional threats that strikes primarily on the most vulnerable. While uniting to the demand for sustainable development as a global and preventative necessity, the articulation of the vulnerable reaffirms the understanding that no national security apparatus has the capacity to handle the security implications of climate change on its own, but builds on the historical responsibility of developed states and the principle of Common but Differentiated Responsibilities, together with a moral responsibility, in legitimatizing the claim for receiving technological and economic aid. The aid is seen to necessitate partnership, principally between states, which provides agency to the vulnerable instead of producing them as helpless and voiceless. While developed states are conceived as vital partners, it is, paradoxically, acknowledged that the societies with predatory consumption patterns, and the structures enabling them, are what prevents the achievement of security from the vulnerable. A similar point is made by Angela Oels (2016), who warns that instead of providing agency to the victims the discursive shift from interventionism to post-interventionism depoliticizes the issue of climate change. According to Oels (ibid., p. 259), where the human security discourse can be criticized for reducing citizens to humans, which denies people's political agency, the post-interventionist discourse fails to problematize the causes of climate change and "rejects the possibility of a geopolitical future in which climate change does not have a devastating impact".

The idea of sustainable development as a technique of security is, however, not accepted by all. Whereas China has in previous years taken a more welcoming and active role in climate security discussions, Russia has maintained its opposing views. In 2018, at a Security Council meeting, Russia expressed a concern over the efforts to link environmental conservation to international peace and security and the potentially misleading and distracting consequences of such actions. Russia reaffirmed its objection toward the role of the Security Council in addressing climate change, while arguing that climate change is an issue that should be dealt within national borders and under the appropriate UN agencies (UN 2018). The logic of traditional security discourse drawing on the idea of (absolute) sovereignty and territorial integrity constructing the heart of international security is prevailing in the statements given by Russia. In this articulation climate change is linked to security only through conflict that is a result of either failing achievement of sustainable development or an insult to the sovereign rights of a state in the name of climate change. The link between sustainable development and security established in the articulation of climate change as a threat multiplier is detached, and climate change is articulated as a sustainable development and not an international security issue. The effects of climate change on human and energy security are separated from international security, and welfare and energy politics are conceived as domestic issues that, only if combined with military power, can be seen as a concern of international security.

However, for the Arctic region the rising discursive shift from traditional to a more integrated and holistic security approach in global security politics can be seen as highly beneficial as the major part of security challenges and threats to the region are nonmilitary by their nature (Luszczuk 2016, p. 37). While the scope of security in the Arctic has already moved from traditional security logic toward inclusion of human and environmental aspects and from confrontation to cooperation, as pointed out by Heininen (2012, 2016), the Arctic states have, simultaneously, shown intentions to rebuild and enhance their Arctic military capabilities (Huebert et al. 2012, p. 18). Environmental and social aspects have a long history in Arctic security discussions where the peace time presence of military and oil production have caused a conflict of interest between Arctic states and their local societies (Heininen 2016, p. 19). While the effects of climate change are being felt in the Arctic both sooner and more dramatically than in most of the other areas, the global climate policies have great impact on the region. The geopolitical ramifications of Arctic sea ice melt and the environmental changes impacting on the well-being and traditional ways of living of the local communities will no doubt have an increasing impact on the security discourses both regionally and globally (see Heininen 2016).

Conclusion: From Reaction to Prevention

Despite the significant increase in the references to climate change in national security strategies and defense planning documents that has put the Security Council under increased pressure to address the climate-related security risks, the Council has been ambivalent in its approach to climate change. Due to the difficult nature of the debates revolving around the dispute over the Council's mandate, climate change is only addressed on a specific-country or conflict-based level. Taking into consideration the concern of the vulnerable states over the national interest-based nature of the Security Council, the inability of the Council to adopt a resolution on climate change signals more about the contention of the traditional hegemonic security structure than to the refusal of the link between climate change and security.

While the concerns over the 'security treatment' of climate change leading to unilateral defensive actions against the vulnerable are nothing but redundant, the analysis suggests that the representations of climate change as a threat are calling for multilateral and preventative measures based on mainstreaming of climate action in the name of security. It seems that rather than security inevitably bringing forth the logic of war and exceptionalism, climate change emphasizes the importance of an integrated and holistic approach in maintaining peace and security. Security is seen to necessitate the recognition of the interconnection and interdependence among states and the symbiotic relationship between human, national and global security. The participation in global governance is taken as a necessity in order to maintain sovereignty that is more threatened by criminal non-state actors than by other states. The risks arising from lacking adaptive capacities of vulnerable social systems need preventative measures that is deeply linked to the promotion of sustainable development. Only if these risk management practices fail are more exceptional security measures called forth.

While the direct threats of climate change such as sea-level rise and extreme weather events have found their way into national security strategies of many states, the security discussion of the Security Council is mostly focused on the indirect threats. Identifying climate change as a threat multiplier is more likely to produce security policies focused on adaptation while the mitigation agenda is largely suppressed. The need for ambitious and effective emission reductions in the name of security is rarely articulated. It, thus, seems that while the effects of climate change are contesting the basic structures of security, which are built on territoriality, sovereignty and military power, are being contested, the idea of peace and security necessitating cumulative economic and social development is mainly left untouched. Security is still predominantly about state survival, but human security is seen increasingly crucial for the survival of the state. The idea of maximizing material power relative to other states is challenged by the idea of the necessity of cooperation and global governance. The interconnectedness of threats that do not respect borders or sectoral boundaries is challenging the image of internal order vs. external disorder, while blurring the limits between private/public, domestic/foreign and between those providing and in need of security. While the meaning of security still remains to be connected to the idea of external anarchy and distinction between human civilization and the biosphere, security is rather conceived in terms of sustaining peace in transformation, than as maintaining any stable state of security.

The Discourse theoretical framework employed in this chapter reveals how in the increasingly accepted articulation of climate change as a threat multiplier, the security signifier is taken out of its conventionalized military context and related to development and humanitarian problems. An equivalential chain is produced between different vulnerabilities of individuals and societies against a threat of chaotic and dangerous future, brought about by climate change. The threat is not simply shifted to the periphery of the social by depersonalizing and differentiating the danger, but articulated with binary framing, where the boundary is drawn between the adaptive and maladaptive societies, poor and rich, responsible and irresponsible or those acting as preventers and the ones becoming dangerous. Vulnerabilities are not only linked to human insecurities, but also to conflicts and forced migration that instead of a reaction necessitates pre-emptive responses that can prevent the potentially chaotic domino effects that put the prevailing international order in danger. By shattering the stability of existing social structures, climate change, in Discourse theoretical terms, can be argued to be inducing the reactivation of deeply sedimented structures.

For the Arctic region, suffering from an accelerated pace of warming, the global climate and security politics are by no means irrelevant. Besides the fact that some Arctic states such as the United States and Russia as great powers have a significant role in global climate and security governance, the local communities living in the Arctic region are one of the first ones facing the geopolitical, environmental and social changes brought about by climate change.

Notes

1. The theory of Laclau and Mouffe is best known from their groundbreaking work published in 1985, Hegemony and Socialist Strategy. Since then the authors themselves and various other scholars have contributed to and reinterpret the theory.
2. In a resolution 1625 of 2005, the Security Council expressed its will and commitment to move beyond a reactive approach to conflict resolution, strengthen its role in conflict prevention, and take action to address the root causes of conflict.
3. Especially, poststructuralist scholars analyzing the discursive shift have particularly drawn on Michel Foucault's notion of biopolitics and analyses of the functioning of different regimes of power, which go beyond the sovereign question of territoriality and centralized legitimate authority. According to Foucault during the eighteenth-century sovereign power started transform into new forms of power—discipline and biopolitics (Aradau and van Munster 2017, p. 163).

References

Aradau, Claudia, and Rens van Munster. 2017. Poststructuralist Approaches to Security. In *Routledge Handbook of Security Studies*, 2nd ed., ed. Myriam Dunn Cavelty and Thierry Balzacq, 157–177. London and New York: Routledge.

Baylis, John. 2014. International and Global Security. In *The Globalization of World Politics: An Introduction to International Relations*, 6th ed., ed. John Baylis, Steve Smith, and Patricia Owens, 229–242. Oxford: Oxford University Press.

Bellamy, Alex. 2010. Humanitarian Intervention. In *Contemporary Security Studies*, 2nd ed., ed. Alan Collins, 359–377. Oxford: Oxford University Press.

Bettini, Giovanni. 2013. (In)convenient Convergences: 'Climate Refugees', Apocalyptic Discourses and the Depoliticization of Climate-Induced Migration. In *In Interpretive Approaches to Global Climate Governance: (De)constructing the Greenhouse*, 1st ed., ed. Chris Methmann, Delf Rothe, and Benjamin Stephan, 201–223. Abingdon: Routledge.

Butler, Judith. 1992. Contingent Foundations: Feminism and the Question of "Postmodernism". In *Feminists Theorize the Political*, ed. Judith Butler and Joan W. Scott. New York: Routledge.

Campbell, David. 1998. *Writing Security*, 2nd ed. Minneapolis: University of Minnesota Press.

Cavelty, Myriam Dunn, and Thierry Balzacq (eds.). 2017. *Routledge Handbook of Security Studies*, 2nd ed. London and New York: Routledge.

Chandler, David. 2012. Resilience and Human Security: The Post-interventionist Paradigm. *Security Dialogue* 43 (3): 213–229.

Chandler, David. 2015. Reconceptualizing International Intervention: Statebuilding, 'Organic Processes' and the Limits of Causal Knowledge. *Journal of Intervention and Statebuilding* 9 (1): 70–88.

Cousins, Stephanie. 2013. UN Security Council: Playing a Role in the International Climate Change Regime? *Global Change, Peace & Security* 25 (2): 191–210.

Crawford, Beverly, and Ronnie Lipschutz. 1997. Discourses of War: Security and the Case of Yugoslavia. In *Critical Security Studies: Concepts and Cases*, ed. Keith Krause and Michel Williams, 149–185. Abingdon, Oxon: Routledge.

Dalby, Simon. 2013. Climate Change: New Dimensions of Environmental Security. *The RUSI Journal* 158 (3): 34–43.

Dalby, Simon. 2015. Climate Geopolitics: Securing the Global Economy. *International Politics* 52: 426–444.

Der Derian, James, and Michael Shapiro (eds.). 1989. *International/Intertextual Relations*. Lexington: Lexington Books.

Detraz, Nicole, and Michele Betsill. 2009. Climate Change and Environmental Security: For Whom the Discourse Shifts. *International Studies Perspectives* 10 (3): 303–320.

Deudney, Daniel. 1990. The Case Against Linking Environmental Degradation and National Security. *Millennium: Journal of International Studies* 19 (3): 461–476.

Dillon, Michael. 1996. *Politics of Security: Towards a Political Philosophy of Continental Thought*. London: Routledge.

Duffield, Mark. 2006. Human Security: Linking Development and Security in an Age of Terror. In *New Interfaces Between Security and Development: Changing Concepts and Approaches*, 11–38. Bonn: German Development Institute, Study No. 13.

Duffield, Mark. 2007. *Development, Security and Unending War: Governing the World of Peoples*. Cambridge: Polity Press.

Duffield, Mark, and Nicholas Waddell. 2006. Securing Humans in a Dangerous World. *International Politics* 43 (1): 1–23.

Elden, Stuart. 2006. Contingent Sovereignty, Territorial Integrity and the Sanctity of Borders. *SAIS Review* XXVI (1): 11–24.

Fidler, David P. 1996. Caught Between Traditions: The Security Council in Philosophical Conundrum. Articles by Maurer Faculty. Paper 745. http://www.repository.law.indiana.edu/facpub/745.

Fierke, Karin. 2015. *Critical Approaches to International Security*. Cambridge: Polity Press.

Gavas, Mikaela, Christine Hackenesch, Svea Koch, James Mackie, and Simon Maxwell. 2016. *The European Union's Global Strategy: Putting Sustainable Development at the Heart of EU External Action*. European Think Tanks Group.

Gilmore, Elisabeth A. 2017. Introduction to Special Issue: Disciplinary Perspectives on Climate Change and Conflict. *Current Climate Change Reports* 3: 193–199. https://doi.org/10.1007/s40641-017-0081-y.

Griggs, Steven, and David Howarth. 2014. Post-structuralism, Social Movements and Citizen Politics. In *Handbook of Political Citizenship and Social Movements*, ed. H.-A. van der Heijden, 279–308. Cheltenham: Edward Elgar.

Hansen, Lene. 2006. *Security as Practice: Discourse Analysis and the Bosnian War*. London: Routledge.

Hartmann, Betsy. 2010. Rethinking Climate Refugees and Climate Conflict: Rhetoric, Reality and the Politics of Policy Discourse. *Journal of International Development* 22 (2): 233–246.

Hartmann, Betsy. 2013. Climate Chains: Neo-Malthusianism, Militarism and Migration. In *Interpretive Approaches to Global Climate Governance (De) Constructing the Greenhouse*, ed. Chris Methmann, Delf Rothe, and Benjamin Stephan, 152–172. Abingdon: Routledge.

Heininen, Lassi. 2012. Security in the Arctic. In *Polar Law Textbook II*, ed. Natalia Loukacheva. Copenhagen: Nordic Council of Ministers.

Heininen, Lassi. 2016. Security of the Global Arctic in Transformation—Potential for Changes in Problem Definition. In *Future Security of the Global Arctic*, ed. Lassi Heininen, 12–34. Basingstoke: Palgrave Macmillan.

Howarth, David. 2000. *Discourse*. Buckingham: Open University Press.

Huebert, R., H. Exner-Pirot, A. Lajeunesse, and J. Gulledge. 2012. *Climate Change & International Security: The Arctic as a Bellwether*. Arlington, VA: Center for Climate and Energy Solutions. Available at http://www.c2es.org/publications/climate-change-international-arctic-security/.

Huysmans, Jef, Andrew Dobson, and Raja Prokhovnik. 2006. *The Politics of Protection: Sites of Insecurity and Political Agency*. London: Routledge.

Jørgensen, Marianne, and Louise Phillips. 2002. *Discourse Analysis as Theory and Method*. London, Thousand Oaks, and New Delhi: Sage.

Käkönen, Jyrki (ed.). 1994. *Green Security or Militarized Environment*. Aldershot: Dartmouth.

Krahmann, Elke. 2003. Conceptualizing Security Governance. *Cooperation and Conflict: Journal of the Nordic International Studies Association* 38 (1): 5–26.

Laclau, Ernesto. 1996. *Emancipation(s)*. London: Verso.

Laclau, Ernesto. 2005. *On Populist Reason*. London and New York: Verso.

Laclau, Ernesto. 2014. *The Rhetorical Foundations of Society*. London: Verso.

Laclau, Ernesto, and Chantal Mouffe. 2001/1985. *Hegemony and Socialist Strategy: Towards a Radical Democratic Politics*. London: Verso.

Lilly, Damian. 2000. *The Privatization of Security and Peacebuilding: A Framework for Action*. London: International Alert.

Luszczuk, Michal. 2016. Military Cooperation and Enhanced Arctic Security in the Context of Climate Change and Growing Global Interest in the Arctic. In *Future Security of the Global Arctic*, ed. Lassi Heininen, 35–54. Basingstoke: Palgrave Macmillan.

Mandel, Robert. 2002. *Armies Without States: The Privatization of Security*. London: Lynne Rinner.

McDonald, Matt. 2002. Human Security and the Construction of Security. *Global Society* 16 (3): 277–295.

McDonald, Matt. 2018. Climate Change and Security: Towards Ecological Security? *International Theory* 10 (2): 153–180.

Mearsheimer, John. 1990. Back to the Future: Instability in Europe After the Cold War. *International Security* 15 (4): 5–56.

Methmann, Chris, and Delf Rothe. 2013. Apocalypse Now! From Exceptional Rhetoric to Risk Management in Global Climate Politics. In *Interpretive Approaches to Global Climate Governance: (De-)Constructing the Greenhouse*, ed. Chris Methmann, Delf Rothe, and Benjamin Stephan, 105–121. London: Routledge.

Nabers, Dirk. 2015. *A Poststructuralist Discourse Theory of Global Politics*. Basingstoke, Hampshire and New York: Palgrave Macmillan.

Oels, Angela. 2012. From 'Securitization' of Climate Change to 'Climatization' of the Security Field: Comparing Three Theoretical Perspectives. In *Climate Change, Human Security and Violent Conflict: Hexagon Series on Human and Environmental Security and Peace*, vol. 8, ed. Jurgen Scheffran, Michael Brzoska, Hans Günter Brauch, Peter Link, and Janpeter Schilling, 185–205. Berlin and Heidelberg: Springer.

Oels, Angela. 2016. Resisting the Climate Security Discourse Restoring "the Political" in Climate Change Politics. In *Reframing Climate Change Constructing Ecological Geopolitics*, ed. Shannon O'Lear and Simon Dalby, 246–265. Abingdon and New York: Routledge.

Palosaari, Teemu. 2012. The Amazing Race: On Resources, Conflict and Cooperation in the Arctic. In *Nordia Geographical Publications Yearbook 2011*, ed. Lassi Heininen and R. Rouge-Oikarinen. Oulu: The Geographical Society of Northern Finland.

Palosaari, Teemu, and Nina Tynkkynen. 2015. Arctic Securitization and Climate Change. In *Handbook of the Politics of the Arctic*, ed. Leif Jensen and Geir Hønneland. Cheltenham and Northampton: Edward Elgar.

Piiparinen, Touko. 2012. Sovereignty-Building: Three Images of Positive Sovereignty Projected Through Responsibility to Protect. *Global Change, Peace & Security* 24 (3): 405–424. https://doi.org/10.1080/14781158.2012.716033.

Rothe, Delf. 2016. *Securitizing Global Warming: A Climate of Complexity.* Abingdon: Routledge.

Scott, Shirley. 2015. Implications of Climate Change for the UN Security Council. *International Affairs* 91 (5): 1317–1333.

Shepherd, Laura, and J. Weldes. 2008. Security: The State (of) Being Free from Danger? In *Globalization and Environmental Challenges: Hexagon Series on Human and Environmental Security and Peace*, vol. 3, ed. H.G. Brauch et al., 529–536. Berlin: Springer.

Stern, Maria, and Joakim Öjendal. 2010. Mapping the Security—Development Nexus: Conflict, Complexity, Cacophony, Convergence? *Security Dialogue* 41 (1): 5–29.

Stripple, Johannes. 2012. The Subject of Security in a Warming World. *The Brown Journal of World Affairs* 18 (2): 181–194.

Torfing, Jacob. 2005. Discourse Theory: Achievements, Arguments, and Challenges. In *Discourse Theory in European Politics Identity, Policy and Governance*, ed. David Howarth and Jacob Torfing, 1–32. Basingstoke, Hampshire and New York: Palgrave Macmillan.

Trombetta, Maria Julia. 2008. Environmental Security and Climate Change: Analysing the Discourse. *Cambridge Review of International Affairs* 21 (4): 585–602.

Weiss, Thomas. 2011. *Thinking About Global Governance: Why People and Ideas Matter.* Abingdon, Oxon: Routledge.

United Nations Documents

UN. April 5, 2007. S/2007/186 Letter Dated 5 April 2007 from the Permanent Representative of the United Kingdom of Great Britain and Northern Ireland to the United Nations Addressed to the President of the Security Council.

UN. July 20, 2011a. S/PV.6587 Security Council Open Debate on Impact of Climate Change.

UN. July 20, 2011b. S/PRST/2011/15 Statement by the President of the Security Council.

UN. July 30, 2015. S/PV.7499 Security Council Open Debate on Peace and Security Challenges Facing Small Island Developing States.

UN. July 11, 2018. S/PV.8307 Security Council Open Debate on Understanding and Addressing Climate-Related Security Risks.

CHAPTER 3

China, Great Power Responsibility and Arctic Security

Sanna Kopra

Abstract Based on the basic premise of the English School theory of International Relations that great powers have a unique responsibility to uphold international peace and security, this chapter examines the ways in which China's notions of *great power responsibility* guide its policies in the Arctic. Thereby the chapter seeks to increase our understanding of what kind of a great power China will be and how its rise may shape the discourses, premises and paradigms of international security in the future. Since there seems to be no real risk of a military conflict in the Arctic, and since China's growing engagement is unlikely to induce such a risk, the chapter will focus on China's contribution to the most imminent security risk in the region: climate change. The chapter demonstrates that although the Chinese government has started to define climate responsibility as an important attribute of great power responsibility at a global level, China's Arctic policy does not make reference to this special responsibility even though climate change generally seems to be one of the key drivers of China's Arctic engagement. Finally, the chapter concludes that because notions of great power responsibility derive

S. Kopra (✉)
Arctic Centre, University of Lapland, Rovaniemi, Finland
e-mail: Sanna.Kopra@ulapland.fi

L. Heininen and H. Exner-Pirot (eds.), *Climate Change and Arctic Security*, https://doi.org/10.1007/978-3-030-20230-9_3

from the practices of security politics, it is not very likely that China—or the United States—will assume a strong climate leadership role without a profound normative transformation in the discourses and premises of international security.

Keywords English School · Great power · Climate responsibility · China's Arctic policy

Introduction

Over the last thirty years, international society has been shaped by two significant changes that are likely fueling a paradigm shift in international security. In particular, climate change has emerged as a key threat of our times. It inevitably poses serious concerns over human security around the world, as it affects the lives of all the people on the planet by increasing the frequency and intensity of extreme weather events, rising sea levels and changing precipitation patterns, for instance. Developing countries—especially the poorest people living in those countries—and small island states in the Pacific and Indian Oceans are most likely to be harmed by the adverse effects of climate change and they do not often have enough resources to adapt to it. In addition, the polar regions are very vulnerable to climate change. In the Arctic, temperatures have risen more rapidly than in any other region on Earth during the last 30 years. As a result, sea ice, permafrost and the snow cover have decreased dramatically, hampering the livelihoods of local people and threatening the survival of many Arctic species such as polar bears, walrus and seals (see, for example, Serreze 2018). Moreover, the melting of ice and permafrost in the High North accelerates climate change and alters ecosystems globally (ACIA 2004; AMAP 2017).

The People's Republic of China (hereafter China or PRC) has undergone a dramatic identity change from an isolated communist state to an (emerging) great power that cannot be ignored in any sector of politics or economic life, including regional affairs in the Arctic. This identity transformation has significantly transformed its sphere of interests and raised global concern about it being a revisionist power seeking to challenge the status quo. Owing to China's growing interest in the Arctic, many regional actors fear that China will challenge the rights and interests of the Arctic states and raise the risk of a military conflict in the region (see, for example, Cassotta et al. 2015; Rainwater 2013; Wright 2011). Instead of directly engaging in this debate, I study the practices of international security from another angle by investigating the ways in which China's notions of *great power responsibility* guide its policies in the Arctic. In this way, I seek to increase our understanding of what kind of a great power China will be and how its rise may shape the discourses, premises and paradigms of international security in the future.

Consequently, my chapter departs from the work of classic realists, who believe that international relations are 'condemned to perpetual great-power competition' (Mearsheimer 2001, p. 2) and hence the rise of a new great power, China, unavoidably causes a hegemonic war. Conversely, I underscore the fact that there is no natural law determining the behaviour of rising powers and that their conduct is influenced by circumstances and ideas instead. In comparison to previous rising powers, China's rise takes place in a very different international architecture that has brought about new global concerns. On the other hand, China's cultural-historic background, ideologies and values influence its priorities in international relations and its behaviour in different situations. I build here on the English School theory of International Relations and start from the fact that a central element of being a great power is to have 'fundamental global capabilities and responsibilities that minor or medium powers do not' (Jackson 2000, p. 21). According to Shambaugh (2013), whether or not China possesses such capabilities and responsibilities at the global level remains to be debated. I argue that we must study *how* China's rise will change the discourses, premises and paradigms of international security, and *why*. Hence, it is no longer meaningful to discuss *whether or not* China's rise to a great power will change the practices of international society. In this chapter, I engage in this debate by elaborating China's policies in the Arctic. Since there seems to be no real risk of a military conflict in the Arctic, and since China's growing engagement is unlikely to induce such a risk, I will focus on China's contribution to the most imminent security risk in the region: climate change.

The basic tenet of English School theory is that states form an *international society* that exists 'when a group of states, conscious of certain common interests and common values, form a society in the sense that they conceive themselves to be bound by a common set of rules in their relations with one another, and share in the working of common institutions' (Bull 2002 [1977], p. 13). In that society, great powers have a special responsibility to ensure its workings and to maintain international peace and security. Yet, the classic English School's focus on the *security of states* is somewhat outdated, as climate change will shape the ways in which security is being conceptualised and assessed in the future. In particular, *human security*, which stresses the protection of human beings from both traditional (i.e. military) and non-traditional security threats (such as poverty, pandemic diseases and climate change), is likely to become a more important approach to security than ever before. In this chapter, therefore, I argue that climate change is a severe security threat that cannot be fully addressed without having a paradigm shift in the future international security discourse. In particular, that shift would yield a profound normative change in international society: while great powers are generally agreed to have a unique responsibility to uphold international peace and security, they would be expected to shoulder the greatest responsibility for climate change mitigation and adaptation to it. For China, this responsibility seems to be an important justification to engage in Arctic affairs and to strive for a more influential role in the regional governance. Therefore, China's Arctic engagement provides a fruitful case to examine its practices of great power responsibility and to analyse their implications for discourses on international security.

Rise of China and Great Power Responsibility

English School theory maintains that power and responsibility go hand in hand: the greater the power of a state is, the greater the international effect of its domestic and foreign policies will be on the workings of international society and the greater its responsibility will be for the collective well-being of the members of that society. As Watson (1982, p. 201) puts it, 'states with enough power to do serious damage to the functioning of international society should accept the responsibility not to cause such damage, but to pursue their interests with prudence and restraint'. In 1945, the United Nations (UN) Charter formally appointed great powers to have special responsibilities in international society: the permanent members of the UN Security Council (China, France, Russia, the United Kingdom, and the United States) have the 'primary responsibility for the maintenance of international peace and security' (Chapter 5, Article 24). This responsibility was largely based on great powers' material capabilities, because, according to then British

Foreign Secretary Anthony Eden, 'the more power and responsibility can be made to correspond, the more likely it is that the machinery will be able to fulfil its functions' (quoted in Bukovansky et al. 2012, p. 31). In practice, great powers must modify their 'policies in the light of the managerial responsibilities they bear' in international society (Bull 2002 [1977], p. 196). As for China's Arctic engagement, this means that China should be cautious not to harm the sovereign rights of Arctic states or otherwise undermine the stability in the region, as such action would be inconsistent with its great managerial power responsibility.

For Watson as well as for most of the other classic English School theorists, great power responsibility means that in a suddenly sharpened conflict, the main responsibility for peace negotiations falls on great powers. In cases of unavoidable confrontation between great powers on opposing sides, the responsibility to avoid an unnecessary resorting to force belongs to them, rather than the 'smaller and more immediate protagonists' (Watson 1982, p. 201). In the post-World War II era, the United States (U.S.) has emphasised human rights and humanitarian intervention as key aspects of great power responsibility, since it views liberal norms as an important means to sustain the workings of international society (Kopra 2019a, b). As China is a signatory to the international human rights treaties and other key international legal frameworks, it respects liberal norms at least in principle without seeking to challenge them. In moral terms, however, China does not accept them as guiding principles of international society and instead seems to seek to transform the ways in which great power responsibility is conceptualised in the future. In these efforts, the international norms of climate responsibility play an important role, as climate change is a truly global phenomenon—not a social construction based on the cultural and philosophical traditions of the West (ibid.).

Within the English School, scholars debate on the justifications and scope of great power responsibility: *pluralists* stress the importance of great powers' role in fostering international order and the security of states, whereas *solidarists* emphasise international justice, human security and the well-being of individuals. Recent contributions to English School theory have also paid attention to the environmental responsibilities of states in general (Jackson 2000; Falkner 2012; Falkner and Buzan 2019), and those of great powers in particular (Kopra 2019a, b). Furthermore, I have argued that *climate responsibility* is an established international norm, and both pluralist and solidarist argumentation can be used to justify great powers' special responsibility for climate change in two respects. First, climate change is a source of potential international conflict, and great powers bear the main responsibility for the resolution of the problem owing to their *functional* great power responsibility to maintain international peace and security. Second, climate change causes severe harm to the well-being of individual human beings around the world, and great powers should lead global endeavours to solve it because of their *diplomatic* great power responsibility to promote international justice and well-being (Kopra 2019a, p. 74).

From the English School standpoint, thus, China's rise to the status of a great power indicates that we can expect it to shoulder more responsibility in international (climate) politics in the future. At the same time, we should notice that China might not be willing to carry out such responsibilities if they do not bring along special *rights* in international society as well. In contrast to other states, great powers enjoy privileges in international society: they have the capability and legal authority to 'play a part in determining issues that affect the peace and security' of international society (Bull 2002 [1977], p. 196). This right is written in the UN Charter: permanent members of the UN Security Council have a right to decide on the necessary means to maintain international peace and security, including the use of military force. Their power of veto in international decision-making gives them a hegemonic position in international society and thus grants them a privilege to defend their interests or to promote particular issues and ideas in that society, for instance. This privilege makes great powers special and establishes an important source of their responsibility in international society (cf. Bukovansky et al. 2012, pp. 31–32). As a permanent member of the UN Security Council and as a rising great power, China should have access to each and every international forum where decisions regarding international peace and security are made—otherwise it would be difficult for it to shoulder great power responsibility in practice. In principle, the Arctic Council is not such a forum, since issues of international security are not on its agenda. However, the situation may change if climate change will be regarded as an international security risk and an important attribute of great power responsibility. At present, great powers do not enjoy any privileges in international climate politics, because they do not carry special responsibilities, either.

That said, there are indeed signs that climate responsibility is becoming an attribute of great power responsibility (Kopra 2019a, b). After the Cold

War, people started to securitize the environment and many studies on environmentally induced conflicts were conducted (for a detailed overview of these developments, see e.g. Trombetta 2008). In 1992, the UN Security Council acknowledged that the 'non-military sources of instability in the economic, social, humanitarian and ecological fields have become threats to peace and security' (UN Security Council 1992). In the mid-2000s, climate change mitigation was also discussed in the context of security politics (Trombetta 2008, pp. 594–595). Since then, the relationship between climate change and violent conflicts has been extensively investigated (i.e. Lee 2009; Mazo 2010; Welzer 2012). These developments induced a debate on the role of the UN Security Council in climate change mitigation: if seen as a risk to international peace and security, climate change can and *should* be added to its agenda (see also Kalliojärvi's chapter in this volume). In 2007, the UN Security Council indeed organised its first discussion on the nexus between climate change, energy and international security, although China and some other states insisted that the Council was not a proper venue for such debate (United Nations 2007). Conversely, Margaret Beckett, the British Foreign Secretary and President of the Council, argued that the UN Security Council should discuss the security impacts of climate change because the 'Council's responsibility was [is] the maintenance of international peace and security, and climate change exacerbated many threats, including conflict and access to energy and food' (ibid.). In 2009, the UN General Assembly called upon relevant UN organizations to tighten up their climate change mitigation efforts and asked the UN Secretary-General to prepare a comprehensive report on the potential security impacts of climate change (UN General Assembly 2009a). The Secretary-General responded by addressing five channels through which climate change could affect international security: vulnerability, development, coping and security, statelessness, and international conflict (UN General Assembly 2009b). In 2011, the UN Security Council again discussed climate security under the German presidency. While the Council failed to issue a formal resolution on climate change or give specific recommendations on how to improve climate security, it issued the first-ever statement on the matter, expressing "its concern that possible adverse effects of climate change may, in the long run, aggravate certain existing threats to international peace and security" (UN Security Council 2011). Two years later, informal talks were held on the topic, but because of the opposition of China and Russia, climate change was not officially defined as an international security risk (Krause-Jackson 2013). In 2018 and 2019, the Security Council debated the nexus between climate change and conflicts again, and the debate is likely to continue in the coming years. Although the UN Security Council has not made a formal resolution on climate change, the fact that it has held several talks on the issue indicates that it has started to be regarded as a security risk, and hence great powers should bear the special responsibility to respond to it (Kopra 2019b, p. 156).

In rhetoric terms, the United States—at least before the term of Donald J. Trump—and China have acknowledged their great power responsibility for climate change. They have stated that their climate cooperation establishes a 'powerful example that can inspire the world' (White House 2013). In particular, President Barack Obama made an explicit link between climate change and great power status in September 2014 by announcing that the U.S. and China 'have a special responsibility to lead' the global efforts to tackle climate change because that is 'what big nations have to do' (Obama 2014). Despite its opposition to hold a formal discussion on climate change at the UN Security Council, China seems to have started to accept the idea that great powers should lead climate change mitigation and adaptation to the problem at the global level. In September 2014, Special Envoy Zhang Gaoli (2014) announced at the UN Climate Summit that 'responding to climate change is what China needs to do to achieve sustainable development at home as well as to fulfil its due international obligation as a responsible major country'. After that, China has made all its key climate commitments public through joint statements with the United States—a choice which indicates that it has made them in reference to its great power status (cf. White House 2014, 2015, 2016a, b). In November 2016, however, the climate change sceptic Donald Trump was elected president, which ended the Sino-American leadership in international climate politics. In June 2017, the United States decided to withdrew from the Paris Climate Agreement established within the United Nations Framework Convention on Climate Change (UNFCCC) in 2015 (White House 2017). Consequently, international society began to expect China, together with the European Union, to step into the shoes of a global climate leader. Chinese President Xi Jinping indeed responded to these expectations in October 2017 by announcing that China will take a 'driving seat' in international climate negotiations (Xi 2017, p. 4).

At present, however, it remains unclear what kind of a climate leader China will be. It continues to stress the historic responsibilities of developed countries and has not indicated a willingness to substantially enhance its own, very moderate climate change mitigation pledges to the UNFCCC, which forms the heart of the Paris Climate Agreement with its reliance on states' nationally determined mitigation strategies. Importantly, China managed to speed up the implementation of developed countries' pre-2020 climate actions: as a result of China's diplomatic efforts, the UN conference in 2017 decided that additional stocktaking sessions regarding progress in emissions reduction and climate finance be held in the following years. For the future of global efforts to tackle climate change, a crucial question is whether China will be using its growing leverage to re-establish the bifurcation of developed and developing countries held back by the Paris Agreement or whether it will implement ambitious low-carbon development plans that halt its emissions growth as soon as possible and inspire the entire international society to implement efficient emissions reductions in time.

China as an Arctic Great Power

In this section, I discuss the construction of China's great power status in the context of Arctic affairs. Although the concept of great power is very vague and the requirements for its recognition are not crystal clear, it has become increasingly evident that China can no longer be excluded from the 'great power club' (Bull 2002 [1977], p. 194; Wight 1999 [1946], p. 42; Kopra 2019a, pp. 70–73). The English School conception of great power integrates the realist and constructivist perspectives: it assumes that "even if a state reaches a certain level of material capacity and has a certain national identity, it does not automatically become a great power, but instead needs to be recognised and accepted by other recognised great powers" (Kopra 2019b, p. 153). As discussed above, responsibility is an important attribute of being a great power from the English School theory point of view. Therefore, I largely focus on China's great power ***identity*** and its notions of great power responsibility in the Arctic and pay less attention to the material aspects of China's great power status, which have been extensively studied elsewhere (see, for example, Brady 2017). That said, we cannot fully understand a state's identity and beliefs without knowing its (material) interests. In China's great power status-building, two 'centenary goals'[1] to be achieved by 2049, the 100th anniversary of the founding of the PRC, no doubt play a central role. Hence, I give a short introduction to China's interests in the Arctic and discuss how the state's Arctic activities seek to advance the construction of its great power status.

China published its long-awaited white paper entitled *China's Arctic Policy* in January 2018 in order to clarify its visions and interests regarding the Arctic—thus decreasing the international speculations and various "China threat theories" (see Deng 2008) that its growing regional role has fuelled. The paper sums up China's Arctic policy goals as follows: 'to understand, protect, develop and participate in the governance of the Arctic, so as to safeguard the common interests of all countries and the international community in the Arctic, and promote sustainable development of the Arctic' (State Council Information Office of the People's Republic of China 2018). For a long time, China's polar interests were chiefly scientific in nature. It signed the Svalbard Treaty in 1925, and Chinese scholars started to conduct research and take part in international research projects in the Arctic in the early 1990s. In 2004, China's first Arctic research station was established in Ny Ålesund, Svalbard, and in 2018, the China-Iceland Arctic Science Observatory began to operate at Karholl, near the town of Akureyri in northern Iceland. China's first home-built icebreaker, *Xuelong II*, will be finished in 2019—well before the United States even starts to build its new heavy icebreaker.[2] In addition to scientific research, economic interests are today an important driver of China's Arctic engagement. According to the U.S. Geological Survey (2008), the 'extensive Arctic continental shelves may constitute the geographically largest unexplored prospective area for petroleum remaining on Earth', and these resources clearly are of interest to China because of its growing hunger for oil, gas and other natural resources that are essential to its economy. To get access to those resources, the Chinese have increased cooperation with transnational corporations and the Arctic states—especially in Russia's Yamal Peninsula, where Chinese partners take part in Yamal LNG, one of the largest liquefied natural gas (LNG) projects in the world. In addition, China 'hopes to work with all parties to build a "Polar Silk Road" through developing the Arctic shipping routes' (State Council Information Office of the People's Republic of China 2018). In particular, Russia's Northern Sea Route is of special interest to China, as it makes it possible to ship LNG from Yamal to China and offers faster and shorter access to the European market.

Growing strategic interests in the Arctic have also increased China's desire to take part in the practices of Arctic regional governance. Being a non-Arctic state, however, China is not eligible to be a full member of the Arctic Council, the key regional forum established in 1996 to enhance cooperation and to coordinate interaction amongst the eight Arctic states and Arctic Indigenous communities on sustainable development and environmental protection in the region. Yet, China got the status of permanent observer in the Arctic Council in 2013, which provides it access to the Council's activities but no right to participate in decision-making. To increase its leverage in regional affairs, China has actively taken part in informal forums of Arctic governance, such as Arctic Frontiers in Tromsø and the Arctic Circle Assembly in Reykjavik. Interestingly, two high-level Chinese representatives spoke at the opening session of the Arctic Circle Assembly in 2017—while American representatives were conspicuous by their absence—and in 2018, the "China Night" event organised by the Chinese Embassy in Iceland "marked another highlight of the assembly" (Ministry of Foreign Affairs of the PRC 2018). Despite their historic tensions, China has also launched a new joint regional initiative with Japan and Korea, the Trilateral High-Level Dialogue on the Arctic, in which the Arctic states can only take part as observers. Moreover, China officially added the Arctic dimension to president Xi Jinping's Belt and Road Initiative in June 2017 (Xinhua 2017). China has also increased science diplomacy and bilateral cooperation with many Arctic states. For example, the China-Nordic Arctic Research Center was established in 2013 and diplomatic relations with the Nordic countries are now promoted via the so-called 5 + 1 diplomacy model (see Sverdrup-Thygeson et al. 2018).

In order to legitimise its growing role in the Arctic, China has started to cultivate an image of a 'near-Arctic state'. According to China's Arctic white paper, China is an 'important stakeholder' as well as an 'active participant, builder and contributor in Arctic affairs who has spared no efforts to contribute its wisdom to the development of the Arctic region' and who 'commits itself to maintaining a peaceful, secure and stable Arctic order' (State Council Information Office of the People's Republic of China 2018). When it comes to international peace and security, the white paper stresses that 'Peace and stability in the Arctic… serves the fundamental interest of all countries including China'. Although the strategy describes China as a 'responsible major country', it does not otherwise make an explicit reference to China's great power status nor to the related great power responsibilities. Yet, it hints that such responsibilities can be attached to China's regional role: 'As a permanent member of the UN Security Council, China shoulders the important mission of jointly promoting peace and security in the Arctic' (ibid.). In line with this, the white paper pledges to commit to existing international law and to respect the sovereign rights of the Arctic states as well as the interests of other states and the broader international society when promoting China's own interests in the region. Overall, the Chinese government constantly underlines China's benevolent role in international affairs, including the Arctic. As for environmental responsibilities, the Arctic white paper assures the world that 'China always gives top priority to resolving global environmental issues, earnestly fulfills [sic] its obligations under relevant treaties, and discharges its responsibility of environmental protection' (ibid.).

While China does not assume great power responsibility in the Arctic, the Arctic white paper does not refer to great power rights, either. This comes as no surprise: such formulation would probably be viewed as threatening rhetoric by the Arctic states and would not support China's pursuit to portray an image of a benevolent actor in the Arctic and beyond. Yet, the white paper makes it very clear that China, an external actor in the region, possesses international rights in the Arctic—and the Arctic states are to respect them. The strategy indeed argues for China's rights to conduct scientific research, navigate, overfly, fish, lay submarine cables and pipelines as well as explore and exploit natural resources in the high seas, to name a few. To justify these international rights in the Arctic region clearly, the Chinese government has adopted a rhetoric strategy that highlights the global dimensions and transformation of the region. In particular, China's Arctic white paper pays surprisingly little attention to the role of the Arctic Council in regional governance and describes the Arctic more or less as a global common. According to the white paper, the 'Arctic situation now goes beyond its original inter-Arctic States or regional nature, having a vital bearing on the interests of States outside the region and the interests of the international community as a whole, as well as on the survival, the development, and the shared future for mankind' (State Council Information Office of the People's Republic of China 2018). In this way, China seems to pursue to advance a discursive shift from a territorial definition of the region towards a more globalised understanding of the region. If successful, this strategy is likely to alter the discourses on international security as well.

China and Climate Security in the Arctic

There is a scientific consensus on climate change being a real global problem exacerbated by human activities, especially by the burning of fossil fuels such as coal and oil. Since greenhouse gas emissions do not respect borders of sovereign states but float throughout the atmosphere, climate change is a global problem that no individual nation-state can solve (or escape) on its own. Therefore, global cooperation and solutions are necessary. Since the early 1990s, international negotiations on climate change have been conducted under the UNFCCC. Owing to its position as the world's greatest carbon emitter and its status of an emerging great power, China now plays a gatekeeper's role in those negotiations. As China's contribution to the UNFCCC has been discussed elsewhere (e.g. Kopra 2019a), in this chapter I will examine the ways in which China's official statements regarding Arctic affairs conceptualize climate change. In particular, I elaborate on whether China's Arctic policies and strategies address great power responsibility for climate change, and if so, how they are used to legitimate China's role in the regional governance.

At present, China's carbon emissions account for about 30% of the global greenhouse gas emissions, which means that China is now the largest carbon emitter in the world. China seeks to halt its carbon emissions growth by 2030 but it remains unclear how much those emissions will grow before the peak. It seemed that the peak was reached between 2014 and 2016, but in 2017 the emissions rose again because of an increasing demand of coal, oil and gas (Climate Action Tracker 2018). Furthermore, China is also a significant source of short-lived climate pollutants (SLCPs)—especially black carbon and methane—that are important contributors to climate change. Approximately 20% of the warming and snow-ice cover loss in the Arctic is attributed to the reduction of the albedo effect (i.e. Earth's ability to reflect sunlight) caused by black carbon emissions (Koch et al. 2011). Yet, the international legal and governance regime steering the reduction of SLCPs in the Arctic is complex and fragmented (Yamineva and Kulovesi 2018), and China's air quality policies have not addressed black carbon per se, probably because of a lack of scientific knowledge on the substance and on the interconnections between air quality, public health and climate policies (Yamineva and Liu 2019).

In essence, China's Arctic policy does not make any new contributions to international climate politics or efforts to reduce black carbon emissions. The white paper acknowledges that China's emissions reduction measures have a 'positive impact on the climatic and ecological environment of the Arctic' but does not pledge China to enhance those actions. Since the strategy seems to take the legitimation of China's Arctic engagement as its ultimate goal, it pays more attention to the nexus between China's domestic climate security and Arctic climate change than to the state's own contribution to the latter. That said, Arctic climate change undoubtedly is an important driver of China's Arctic engagement, as recent scientific findings show that the melting of Arctic icecaps increases haze pollution in eastern China (Wang et al. 2015); causes flooding in many of China's coastal mega-cities including Shanghai, Tianjin and Hong Kong; and alters many global natural processes that may hinder agricultural production in China. Air pollution and food security are of significant interest to the Chinese party-state because they have a great impact on social stability in the country and thereby on the legitimacy of the Chinese Communist Party itself. Therefore, the Arctic white paper stresses the importance of scientific research on Arctic climate change and its global ramifications, and China wants to learn more about its domestic (security) implications. This knowledge may help improve its climate change adaptation plans, which is highly relevant from the viewpoint of Chinese leadership's domestic responsibility, given the poor preparedness of the entire Chinese society to deal with risks posed by climate change (see China's National Development and Reform Commission 2013).

In terms of global climate leadership, however, China's Arctic white paper misses a great opportunity to convince the world of the state's commitment to climate change mitigation and to making climate responsibility an important attribute of great power responsibility. Since China is the greatest carbon emitter in the world, it is also the most significant contributor to Arctic climate change. When China's Arctic white paper was published, an established (Arctic) great power, the United States, had abandoned the Paris climate agreement, making the future of international climate negotiations look rather grim. Since Russia as another Arctic great power has not showed much willingness to tackle climate change either, China could have rather easily elevated its profile in international society by introducing new, ambitious measures to mitigate climate change. At present, China's nationally determined contribution to the UNFCCC[3] is 'highly insufficient' to reach the goals of the Paris climate agreement unless other states increase their level of ambition (Climate Action Tracker 2018). The Intergovernmental Panel on Climate Change (IPCC) has made it very clear that to prevent the most dangerous climate change from happening, we must reduce the global net carbon emissions by 45% from the 2010 level by 2030 (IPCC 2018). Hence, all states should upgrade their climate mitigation pledges, as they are currently not sufficient to limit the global temperature rise to 2 °C, not to mention 1.5 °C that the IPPC defines as a safer upper limit to global warming. Like many other countries, China is not very eager to commit itself to quantitative emission reduction targets at a global level, since it would 'lose face' if it failed to meet them. Therefore, the bottom-up approach of the Paris climate agreement is particularly appealing to China. In political terms, however, China's Arctic white paper could have shown global leadership by announcing new domestic measures to mitigate climate change for the protection of the Arctic. Such an announcement would have demonstrated that China has truly taken the driver's seat in international climate politics. Besides, there are strong domestic incentives to decrease the use of coal, for instance, because of a growing discontent over environmental pollution amongst Chinese citizens. Moreover, such an announcement might have increased the willingness of the wider international community to recognize China as a legitimate stakeholder in the Arctic and beyond.

Conclusions

Overall, Arctic engagement plays an important role in China's efforts to become an established great power in terms of economy, technology, knowledge and policy. Although China is a non-Arctic state, it has found ways to increase its leverage in the domains of economy, science diplomacy and bilateral cooperation with several Arctic states, especially Russia and the Nordic countries. As China's military and technological capabilities have been discussed in previous literature, I chose not to discuss the material aspects of China's great power status in this chapter. Instead, I focused on China's great power identity and corresponding notions of responsibility in the Arctic in general, and climate responsibility in particular. Climate change is indeed a central issue in view of China's increasing role in the Arctic: China argues that because of the adverse effects of climate change, it has special interests in the Arctic and must therefore have a chance to engage in Arctic governance. In other words, China uses climate change as a key justification for its Arctic engagement.

There is a growing consensus in international society that climate change is a threat in terms of both traditional and non-traditional definitions of security. From the viewpoint of English School theory, it means that great powers must shoulder a special responsibility in responding to climate change. Although China has sought to define climate responsibility as an attribute of great power responsibility at a global level (Kopra 2019a), its Arctic policy makes no reference to this special responsibility nor demonstrates any kind of leadership in global efforts to tackle climate change—an omission very critical to international (climate) security. Under the presidency of Donald Trump, the United States has also failed to live up to its great power climate responsibility. If the next U.S. president does not renew American great power leadership in international climate politics, China's emerging climate leadership becomes even more critical in the future. For China, there are strong domestic incentives to take the findings of the IPCC (2018) seriously. In addition to national security interests, the "greenification" of economy supports China's structural reforms and development plans. Thus, taking a stronger leadership role in international climate politics would probably not be a big sacrifice for China. Conversely, such a leadership role would significantly improve its status as a responsible great power—a status that would bring along privileges in international society, perhaps including more room to manoeuvre in the Arctic. Since notions of great power responsibility derive from practices of security politics, however, it is not very likely that either China or the United States will assume a strong climate leadership role without a profound transformation in the discourses and premises of international security. What we need, hence, is a paradigm shift in international security: climate change must be identified as a key security threat of our times. In practice, it would mean that great powers are no longer able to carry out their special responsibility to maintain international peace and security without taking a lead in international efforts to prevent dangerous climate change from happening.

Acknowledgements The author wishes to thank the Academy of Finland (project no. 315402) for funding this research, and the editors as well as Karoliina Hurri for their constructive feedback on the earlier versions of this chapter.

Notes

1. The two centenary goals were originally introduced by the then-President Jiang Zemin at the 15th National Congress of the Chinese Communist Party in 1997. However, the current president, Xi Jinping, elevated them to being China's strategic priorities. They read as follows: (1) 'doubling the 2010 GDP and per capita income of urban and rural residents and finishing the building of a society of initial prosperity in all respects' and (2)

'turning China into a modern socialist country that is prosperous, strong, democratic, culturally advanced and harmonious' by 2049 (Xinhua 2014).

2. For further information on China's (scientific) presence in the Arctic, see Brady (2017).
3. China's current nationally determined contribution to the UNFCCC reads as follows: (a) To achieve the peaking of carbon dioxide emissions around 2030 and making best efforts to peak early; (b) To lower carbon dioxide emissions per unit of GDP by 60–65% from the 2005 level; (c) To increase the share of non-fossil fuels in primary energy consumption to around 20%; and (d) To increase the forest stock volume by around 4.5 billion cubic meters on the 2005 level (China's National Development and Reform Commission 2015).

References

ACIA. 2004. Impacts of a Warming Arctic: Arctic Climate Impact Assessment. http://amap.no/acia/.

AMAP. 2017. *Snow, Water, Ice and Permafrost in the Arctic (SWIPA) 2017*. Arctic Monitoring and Assessment Programme. https://www.amap.no/documents/doc/Snow-Water-Ice-and-Permafrost-in-the-Arctic-SWIPA-2017/1610.

Brady, Anne-Marie. 2017. *China as a Polar Great Power*. Cambridge: Cambridge University Press.

Bukovansky, Mlada, Ian Clark, Robyn Eckersley, Richard Price, Christian Reus-Smit, and Nicholas J. Wheeler. 2012. *Special Responsibilities: Global Problems and American Power*. New York: Cambridge University Press.

Bull, Hedley. 2002 [1977]. *The Anarchical Society: A Study of Order in World Politics*, 3rd ed. Basingstoke: MacMillan Press.

Cassotta, S., K. Hossain, J.Z. Ren, and M.E. Goodsite. 2015. Climate Change and China as a Global Emerging Regulatory Sea Power in the Arctic Ocean: Is China a Threat for Arctic Ocean Security? *Beijing Law Review* 6: 199–207. http://dx.doi.org/10.4236/blr.2015.63020.

China's National Development and Reform Commission. 2013. 国家适应气候变化战略 [National Climate Change Adaptation Strategy]. http://www.sdpc.gov.cn/zcfb/zcfbtz/201312/W020131209343322750059.pdf.

China's National Development and Reform Commission. 2015. Enhanced Action on Climate Change: China's Intended Nationally Determined Contributions. http://www4.unfccc.int/ndcregistry/PublishedDocuments/China%20First/China%27s%20First%20NDC%20Submission.pdf.

Climate Action Tracker. 2018. China. https://climateactiontracker.org/countries/china/.

Deng, Y. 2008. *China's Struggle for Status: The Realignment of International Relations*. Cambridge: Cambridge University Press.

Falkner, Robert. 2012. Global Environmentalism and the Greening of International Society. *International Affairs* 88 (3): 503–522.

Falkner, Robert, and Barry Buzan. 2019. The Emergence of Environmental Stewardship as a Primary Institution of Global International Society. *European Journal of International Relations* 25 (1): 131–155.

IPCC. 2018. *Global Warming of 1.5°C: An IPCC Special Report on the Impacts of Global Warming of 1.5°C Above Pre-industrial Levels and Related Global Greenhouse Gas Emission Pathways, in the Context of Strengthening the Global Response to the Threat of Climate Change, Sustainable Development, and Efforts to Eradicate Poverty*. Geneva, Switzerland: World Meteorological Organization.

Jackson, Robert H. 2000. *The Global Covenant*. Oxford: Oxford University Press.

Koch, D., S.E. Bauer, A. Del Genio, G. Faluvegi, J.R. McConnell, S. Menon, R.L. Miller, D. Rind, R. Ruedy, G.A. Schmidt, and D. Shindell. 2011. Coupled Aerosol-Chemistry-Climate Twentieth-Century Transient Model Investigation: Trends in Short-Lived Species and Climate Responses. *Journal of Climate* 24: 2693–2714.

Kopra, Sanna. 2019a. *China and Great Power Responsibility for Climate Change*. London and New York: Routledge.

Kopra, Sanna. 2019b. China, Great Power Management, and Climate Change: Negotiating Great Power Climate Responsibility in the UN. In *International Organization in the Anarchical Society: The Institutional Structure of World Order*, ed. Tonny Brems Knudsen and Cornelia Navari. New York: Palgrave Macmillan.

Krause-Jackson, Flavia. 2013. Climate Change's Links to Conflict Draws UN Attention. *Bloomberg*, February 15. http://www.bloomberg.com/news/articles/2013-02-15/climate-change-s-links-to-conflict-draws-un-attention.

Lee, James R. 2009. *Climate Change and Armed Conflict: Hot and Cold Wars*. London: Routledge.

Mazo, Jeffery. 2010. *Climate Conflict: How Global Warming Threatens Security and What to Do About It*. London: Routledge.

Mearsheimer, J.J. 2001. *The Tragedy of Great Power Politics*. New York: W. W. Norton.

Ministry of Foreign Affairs of the PRC. 2018. China Night Lights Up 2018 Arctic Circle Assembly. https://www.fmprc.gov.cn/mfa_eng/wjb_663304/zwjg_665342/zwbd_665378/t1606536.shtml.

Obama, Barack. 2014. Remarks by the President at U.N. Climate Change Summit. https://obamawhitehouse.archives.gov/the-press-office/2014/09/23/remarks-president-un-climate-change-summit.

Rainwater, S. 2013. Race to the North: China's Arctic Strategy and Its Implications. *Naval War College Review* 66 (2).

Serreze, Marc C. 2018. *Brave New Arctic: The Untold Story of the Melting North*. Princeton: Princeton University Press.

Shambaugh, David. 2013. *China Goes Global: The Partial Power*. New York: Oxford University Press.

State Council Information Office of the People's Republic of China. 2018. China's Arctic Policy. http://english.gov.cn/archive/white_paper/2018/01/26/content_281476026660336.htm.

Sverdrup-Thygeson, Bjørnar, Wrenn Yennie Lindgren, and Marc Lanteigne (eds.). 2018. *China and Nordic Diplomacy*. London and New York: Routledge.

Trombetta, Maria Julia. 2008. Environmental Security and Climate Change: Analysing the Discourse. *Cambridge Review of International Affairs* 21 (4): 585–602.

United Nations. 2007. Security Council Holds First-Ever Debate on Impact of Climate Change on Peace, Surety, Hearing Over 50 Speakers. http://www.un.org/press/en/2007/sc9000.doc.htm.

UN General Assembly. 2009a. Climate Change and Its Possible Security Implications. A/RES/63/281, 85th Plenary Meeting, June 3.

UN General Assembly. 2009b. Climate Change and Its Possible Security Implications. Report of the Secretary-General. A/64/350, September 11.

UN Security Council. 1992. Note by the President of the Security Council. S/23500, January 31.

UN Security Council. 2011. Statement by the President of the Security Council. S/PRST/2011/15, July 20.

U.S. Geological Survey. 2008. Circum-Arctic Resource Appraisal: Estimates of Undiscovered Oil and Gas North of the Arctic Circle. http://pubs.usgs.gov/fs/2008/3049/fs2008-3049.pdf.

Wang, Hui-Jun, Huo-Po Chen, and Liu Jiping. 2015. Arctic Sea Ice Decline Intensified Haze Pollution in Eastern China. *Atmospheric and Oceanic Science Letters* 8 (1): 1–9.

Watson, Adam. 1982. *Diplomacy: The Dialogue Between States*. London: Eyre Methuen.

Welzer, Harald. 2012. *Climate Wars: Why People Will Be Killed in the Twenty-First Century*, trans. Patrick Camiller. Cambridge: Polity Press.

White House. 2013. Joint U.S.-China Statement on Climate Change. https://2009-2017.state.gov/r/pa/prs/ps/2013/04/207465.htm.

White House. 2014. U.S.-China Joint Announcement on Climate Change. https://www.whitehouse.gov/the-press-office/2014/11/11/us-china-joint-announcement-climate-change.

White House. 2015. U.S.-China Joint Presidential Statement on Climate Change. https://obamawhitehouse.archives.gov/the-press-office/2015/09/25/us-china-joint-presidential-statement-climate-change.

White House. 2016a. U.S.-China Joint Presidential Statement on Climate Change. https://obamawhitehouse.archives.gov/the-press-office/2016/03/31/us-china-joint-presidential-statement-climate-change.

White House. 2016b. U.S.-China Climate Change Cooperation Outcomes. https://obamawhitehouse.archives.gov/the-press-office/2016/09/03/fact-sheet-us-china-cooperation-climate-change.

White House. 2017. Statement by President Trump on the Paris Climate Accord. https://www.whitehouse.gov/briefings-statements/statement-president-trump-paris-climate-accord/.

Wight, Martin. 1999 [1946]. *Power Politics*. London: Leicester University Press.

Wright, C.D. 2011. *The Dragon Eyes the Top of the World: Arctic Policy Debate and Discussion in China*. China Maritime Institute, U.S. Naval War College. http://www.usnwc.edu/Research—Gaming/China-Maritime-Studies-Institute/Publications/documents/China-Maritime-Study-8_The-Dragon-Eyes-the-Top-of-.pdf.

Xi, Jinping. 2017. Secure a Decisive Victory in Building a Moderately Prosperous Society in All Respects and Strive for the Great Success of Socialism with Chinese Characteristics for a New Era. http://www.xinhuanet.com/english/download/Xi_Jinping's_report_at_19th_CPC_National_Congress.pdf.

Xinhua. 2014. Xi Eyes More Enabling Int'l Environment for China's Peaceful Development. http://en.people.cn/n/2014/1130/c90883-8815967.html.

Xinhua. 2017. Vision for Maritime Cooperation Under the Belt and Road Initiative. http://www.xinhuanet.com/english/2017-06/20/c_136380414.htm.

Yamineva, Yulia, and Kati Kulovesi. 2018. Keeping the Arctic White: The Legal and Governance Landscape for Reducing Short-Lived Climate Pollutants in the Arctic Region. *Transnational Environmental Law* 7 (2): 201–227.

Yamineva, Yulia, and Zhe Liu. 2019. Cleaning the air, Protecting the Climate: Policy, Legal and Institutional Nexus to Reduce Black Carbon Emissions in China. *Environmental Science & Policy* 95: 1–10.

Zhang, Gaoli. 2014. Build Consensus and Implement Actions for a Cooperative and Win-Win Global Climate Governance System. http://www.fmprc.gov.cn/mfa_eng/zxxx_662805/t1194637.shtml.

CHAPTER 4

Climate Change Ethics in the Arctic

Teemu Palosaari

Abstract A significant share of the world's as-yet unexploited oil and gas resources are at the bottom of the Arctic Ocean. As the sea ice is melting, coastal states and energy companies are viewing these resources with interest. Using them would, however, would accelerate climate change, leading to a heated debate on whether the new Arctic oil and gas reserves should be utilized or left untouched. This chapter's main argument is that it is becoming increasingly difficult for governments, businesses and decision-makers to ignore climate ethics in the Arctic. The way that problems and solutions regarding Arctic oil and gas are defined and promoted has global significance. Arctic glacier and sea-ice melt have both symbolic value in global climate politics as well as a direct impact on the rise in sea level around the world.

Keywords Ethics · Oil and gas · Sustainability · Global climate change · Indigenous peoples

A significant share of the world's as-yet unexploited oil and gas resources are at the bottom of the Arctic Ocean. As the sea ice is melting, coastal

T. Palosaari (✉)
Tampere, Finland
e-mail: teemu.palosaari@icloud.com

L. Heininen and H. Exner-Pirot (eds.), *Climate Change and Arctic Security*, https://doi.org/10.1007/978-3-030-20230-9_4

states and energy companies view these northern resources with great interest. Using them would, however, cause emissions and accelerate climate change. A heated debate has started on whether the new Arctic oil and gas reserves should be utilized or left untouched. This essay looks at how the key questions of global climate change ethics—e.g. moral responsibility, distribution of benefits and burdens, and environmental justice (Wapner 2014; Mittler 2014)—appear in the Arctic discourse. The main argument is that it is becoming increasingly difficult for governments, businesses and decision-makers to ignore climate ethics in the Arctic. The way that problems and solutions regarding Arctic oil and gas are defined and promoted has global significance. Arctic glacier and sea-ice melt has a huge symbolic value in global climate politics as well as a direct impact on the rise in sea level around the world.

The good news first. Despite the vast untapped oil and gas reserves in the seabed, the Arctic has remained peaceful. So, no "race to resources" and no "Cold Rush," let alone a "new Cold War." These phrases make catchy headlines in the news, but, among Arctic scholars, the mainstream view has been that international cooperation between the Arctic Ocean coastal countries—Canada, Iceland, Norway, Russia, the United States and Denmark/Greenland—is working well, and the Arctic is likely to remain peaceful in the near future (Palosaari 2012).

Indeed, the Arctic is an exemplary case where international law actually seems to matter. Coastal countries have been mapping the seabed in order to hand scientific evidence to the United Nations. The UN Commission on the Convention on the Continental Shelf will then make recommendations regarding maritime boundaries. The ownership of underwater minerals, oil and gas will be determined by international law, not by flag-planting. The Arctic coastal countries have jointly declared that they will follow the UN Convention on the Law of the Sea. It appears to be in the national interest of these countries, and in the economic interest of energy companies, to keep the peace in this promising oil and gas production region.

The bad news, then? Climate change is hitting the Arctic hard. The average temperature has risen twice as fast in the Arctic as in most of other regions. The summer ice cover reached an all-time low in 2016. The Greenland glaciers are melting faster than many scientists expected. Climate change has opened access to fossil fuel reserves previously hidden under the ice. Hence, an Arctic paradox: we are getting more oil and gas by burning it. The faster we use fossil fuels, the sooner we gain access to new resources. This is because fossil fuels contribute to climate warming, which melts the Arctic sea ice, making new oil and gas resources accessible. Using those resources would further accelerate climate warming (Palosaari 2012). Therefore, Arctic oil and gas exploration in the areas historically covered by sea ice is inescapably linked to global climate change ethics: Is it acceptable to exploit new oil and gas in the Arctic at a time when humankind needs to reduce carbon dioxide emissions?

Until very recently, the dominant view has been that increased Arctic oil and gas development is inevitable, like it or not. The tempting numbers 13 and 30 appear frequently in speeches of Arctic politicians and businessmen, meaning that the Arctic accounts for about 13% of the world's undiscovered and technically recoverable oil and 30% of gas. Those are powerful numbers. However, a new number, 2 °C, has surfaced in Arctic discussions, thanks to global climate negotiations and the recent Paris Treaty aimed at keeping the increase of global average temperature to "well below 2°C" above the pre-industrial levels. It has been calculated that in order to limit the average global warming to 2 °C, globally a third of oil reserves, half of gas reserves and over 80% of current coal reserves should remain unused from 2010 to 2050 (McGlade and Ekins 2015). A conclusion drawn in the same study, published in *Nature* January 2015, was that there is a contradiction between Arctic oil and gas development and the global warming target. Consequently, the numbers 13/30 and 2 appear to be mutually exclusive. They are like two North Poles of magnets: try to bring them into contact with each other, and they repel each other. Exploitation of the Arctic 13/30 means pushing away the 2 °C target.

Indeed, key issues in the global climate change ethics debate have lately found their way into the speeches of Arctic politicians, businesses, scientific institutions, and environmental organizations. In light of the major annual Arctic meetings of politicians, business, academia and NGOs (e.g. the Arctic Circle Assembly held in Iceland and the Arctic Frontiers Conference in Norway), there are two competing main ways to perceive and communicate about Arctic oil and gas. They build on different ethical undertones and differ in how they define the main problem and moral responsibility and what recommended action they promote.

A key dividing line is whether the main problem regarding Arctic oil and gas development is framed as a regional environmental problem or a global climate issue. The regional approach highlights the risk of oil leaks and contamination from extraction and transportation. Arctic flora and fauna are especially vulnerable, since oil breaks down more slowly under cold, dark conditions, and plants and animals in the Arctic need more time to recover. Seen this way, Arctic oil and gas development becomes a limited environmental problem that can be solved with tools that are readily available: environmental protection and monitoring, scientific standards and advanced extraction technology. The appeal of this framing is boosted by the fact that environmental cooperation has been a key theme in Arctic cooperation since the late 1980s (Mikkelsen and Langhelle 2008). The catchphrase "the vulnerable Arctic" is deeply engrained in the Arctic discourse. It resonates with the successful Arctic intergovernmental collaboration on marine protection and conservation. Sustainable development and the need for "highest environmental standards" appear frequently in the speeches of governments, energy companies, financing institutions and many other stakeholders when they are discussing the economic potential of Arctic oil and gas.

The conclusion from this viewpoint is typically that the use of new Arctic fossil fuel resources is ethically justifiable since it can be done sustainably and in an environmentally-friendly way. It builds on ecological modernization thinking in that it shares the belief that human knowledge and technology can overcome environmental obstacles and that scientific solutions offer a way to sustainable economic growth (Mikkelsen and Langhelle 2008). It emphasizes the human aspect of sustainable development and argues that Arctic oil and gas extraction is imperative for local socio-economic well-being because it can provide work opportunities and welfare.

Presenting Arctic oil and gas development as a problem to be solved by Arctic environmental protection and sustainable extraction has been challenged by a more global viewpoint on the Arctic. This has meant moving the attention from vulnerable Arctic nature to the global greenhouse gas emissions that utilization of the new under-ice oil and gas reserves would result in. The key argument is that since oil and gas have a climate impact irrespective of their geographical place of origin, it is questionable to promote further oil and gas development in the Arctic. The interpretation of sustainable development is different in that it highlights environmental sustainability and argues that, due to the resultant emissions, Arctic oil and gas development cannot be a sustainable use of natural resources, no matter how environmentally friendly and safe the oil and gas extraction and transportation might be.

In the current global climate change ethics debate, a fundamental issue is the responsibility for climate change mitigation, adaptation and compensation (Arnold 2014). In their speeches, heads of the Arctic countries typically handle the moral responsibility issue by arguing that climate change is a global challenge that requires a global response. This is a way of saying that solving climate change is a global rather than Arctic responsibility. The conclusion regarding Arctic oil and gas development is, then, that there is no special "Arctic" responsibility to refrain from using the new resources. Sometimes the causal connection between climate responsibility and Arctic oil and gas is bypassed by arguing that it is the consumption of fossil fuels, not their production, that matters. Emissions are related to the use of oil and gas, and that mostly takes place outside the Arctic. The moral responsibility is discussed, but eventually outsourced to the global level. What facilitates such framing is that a voluntary self-marginalization evoking the region's peripheral status is still readily available in the discourse on the Arctic, despite recent globalization trends. A marginal, remote region cannot be expected to solve global problems alone.

Yet, a causal connection between Arctic resources and global climate change appears more and more often in the Arctic discourse. This includes pointing out the link between the rise in sea level and the melting of glaciers and sea ice. The conclusion is that, since the Arctic produces energy resources that cause greenhouse gas emissions, the Arctic actors economically benefiting from that cannot escape the moral responsibility. Often the arguments reflect the displacement concept that has been much used in global climate ethics. In the Arctic context it means enjoying the benefits of fossil fuels now and moving the harms of climate change to future generations or to impoverished people who have little say in global politics. The displacement of harms from humans to other species is also observable in the discussions, and the iconic polar bear is often mentioned in this light.

However, Indigenous peoples' representatives stress that, while some of the causes of climate change are in the Arctic, the people who have inhabited the Arctic for thousands of years are not to blame. (Numerous Indigenous peoples live in the Arctic: the Inuit, Saami, Aleut, and Nenets, for instance.) They often refer to the industrialized countries' way of life and how they cause pollution and emissions, telling them to look at their own backyard before setting sight on the homelands of Indigenous peoples. Nevertheless, the social aspect of sustainable development is highlighted in their speeches: the Arctic Indigenous peoples should not be denied their equal right to development and economic growth. Sometimes they interpret the moral and environmental climate concerns regarding Arctic fossil resources as neo-colonialism or "mummification": the developed world has used fossil fuels for its economic growth and progress, but Indigenous peoples are denied the right to do so. Instead, they are asked to remain as a sort of outdoor museum frozen in time. Yet, the sustainable use of natural resources is a central aspect of Arctic Indigenous cultures, and this aspect remains visible in the arguments of their representatives.

On the other hand, the link between Arctic oil and gas development and local economic growth has also been questioned. From the viewpoint of Indigenous peoples and other Arctic populations (it is estimated that just ten percent of the total Arctic population consists of Indigenous peoples), the risk is that they will be left with deteriorating living conditions, pollution, coastal erosion and relocated villages, while the revenues travel south, to the capitals of Arctic countries. What happens, for instance, to traditional livelihoods such as hunting, sealing and fishing when the sea ice melts? What would increasing participation in Arctic oil and gas development mean to the Indigenous relationship between land, nature and man? What opportunities do local populations have to exert any kind of influence on Arctic issues when it is becoming a hot spot in global resource politics? Are the local benefits worth the local cost? (Young 2013; Smith 2011).

So, to drill or not to drill, that is the question. Obviously, the above-presented two main views also come up with differing recommended courses of action. The one highlighting sustainable development and global responsibility for climate mitigation ends up promoting Arctic oil and gas development. The conclusion is that further Arctic oil and gas development is acceptable because the related environmental risks can be solved sustainably and because climate change is not the responsibility of the Arctic, but rather a global issue requiring global solutions. Furthermore, it is argued that the Arctic region, including its indigenous peoples, has a right to economic growth. Such a perception has been the dominant one in the Arctic discourse, but is currently being challenged by an Arctic application of global climate change ethics. The different problem definition, causal interpretation and moral evaluation result in promoting a drilling ban on Arctic oil and gas historically located under the sea ice, or at least strict quotas for oil and gas production. At the same time, it has meant highlighting the Arctic renewable energy potential and, for instance, the successful examples of Icelandic geothermal energy and Norwegian hydropower. Furthermore, new topics have entered the discussion, such as "decarbonizing" the Arctic and turning the Arctic into a symbol and leader in the global fight against climate change.

In some cases, new economic terms and concepts such as "divestment," "stranded assets" and "carbon bubble" have appeared in the Arctic debate. A few point out how some international investors are turning away from companies involved in extracting fossil fuels because of the economic risks involved. The economic potential of Arctic oil and gas development is also questioned by others noting that it is likely to be more expensive than elsewhere due to the harsh environmental conditions. Additionally, the application of carbon budget argument in the Arctic claims that new Arctic reserves simply do not fit into the amount of fossil fuels that can be burned without reaching the 2 °C limit.

Despite the different views on urgency and responsibility in the Arctic discourse, it is becoming clear that the ethical issues and global implications can no longer be ignored. The issue of climate justice has found its way to the Arctic. Yet, as in global climate governance, there is a risk of "shadow solutions" in the Arctic: solutions to climate change that, in the words of Stephen Gardiner, "not only fail to address the core features of the problem at hand but also create the dangerous illusion of genuine action" (Gardiner 2014). Particularly when sustainable development is recommended as the solution, there is a risk that, instead of discussing the emissions resulting from Arctic oil and gas, the attention is turned to very specific solutions—such as Arctic offshore oil and gas guidelines—that could help the vulnerable Arctic environment and prevent oil spills or discharge from ships. Although tackling the environmental risks of extraction and transport is important, this viewpoint fails to recognize the climate impact of Arctic oil and gas and related ethical issues.

The way fossil fuels and climate change are tackled in the Arctic will have a broad significance. Since temperatures are rising faster in the Arctic than anywhere else and the consequences are highly visible, Arctic climate governance has a huge symbolic value. Furthermore, climate change in the Arctic has significant environmental implications elsewhere: the sea level rise linked to the melting Greenland glaciers and Arctic Ocean sea ice is already being felt in the Southern hemisphere, e.g. in Bangladesh and the small island states in the Pacific.

References

Arnold, Denis G. (ed.). 2014. *The Ethics of Global Climate Change*. Cambridge: Cambridge University Press.

Gardiner, Stephen M. 2014. *A Perfect Moral Storm: The Ethical Tragedy of Climate Change*. Oxford: Oxford University Press.

McGlade, Christopher, and Paul Ekins. 2015. The Geographical Distribution of Fossil Fuels Unused When Limiting Global Warming to 2°. *Nature* 517: 187–190.

Mikkelsen, Aslaug, and Oluf Langhelle (eds.). 2008. *Arctic Oil and Gas: Sustainability at Risk?* Abingdon: Routledge.

Mittler, Daniel. 2014. The Changing Ethics of Climate Change. *Ethics & International Affairs* 28: 351–358.

Palosaari, Teemu. 2012. The Amazing Race: On Resources, Conflict, and Cooperation in the Arctic. *Nordia Geographical Publications* 40: 13–30. Open access http://tampub.uta.fi/handle/10024/67984.

Smith, Laurence C. 2011. *The New North: The World in 2050*. London: Profile Books.

Wapner, Paul. 2014. Ethical Enhancement in an Age of Climate Change. *Ethics & International Affairs* 28: 325–334.

Young, Oran. 2013. Arctic Stewardship: Maintaining Regional Resilience in an Era of Global Change. *Ethics & International Affairs*, January 7. Available at http://www.ethicsandinternationalaffairs.org/2013/arctic-stewardship-maintaining-regional-resilience-in-an-era-of-global-change/. Last accessed 17 Nov 2016.

CHAPTER 5

Cities and Human Security in a Warming Arctic

Wilfrid Greaves

Abstract In few places around the world are the issues of climate change, urbanization, and security more salient than in the circumpolar Arctic region. As a highly and increasingly urbanized global region that is also experiencing the most acute and widespread effects of global climate change, the Arctic must confront myriad public policy questions implicated by the growth of its cities under conditions of dramatic regional warming. In this chapter, I argue that cities are paradoxical with respect to providing human security for their populations in the face of environmental change, simultaneously reducing individuals' environmental impact and insulating them from certain environmental hazards while making them reliant upon unsustainable and potentially vulnerable economic and social systems that both contribute to the production of environmental threats and expose individuals to risks of catastrophic harm. I outline three reasons to doubt whether Arctic cities will be able to protect the human security of their residents under conditions of environmental change: (1) Arctic cities are vulnerable to the effects of climate change in ways that undermine their critical infrastructure; (2) urbanization

W. Greaves (✉)
University of Victoria, Victoria, Canada
e-mail: wgreaves@uvic.ca

L. Heininen and H. Exner-Pirot (eds.), *Climate Change and Arctic Security*, https://doi.org/10.1007/978-3-030-20230-9_5

significantly increases local warming, meaning that the growth of Arctic cities will generate a positive feedback loop that will worsen the impacts of climate change; and (3) many Arctic cities rely on fossil fuel-based industries, producing a paradox whereby they are simultaneously threatened by and reliant upon the continuation of the very economic activities responsible for contributing to global climate change.

Keywords Cities · Urbanization · Climate change · Infrastructure · Human security · Environmental change

Introduction

Humanity has become an urbanized species, with a majority of people around the world living in urban areas. While levels of urbanization still vary considerably across regions, cities have become the essential sites of human existence in the twenty-first century, and are thus implicated in a host of social, political, and economic questions ranging far beyond the traditional boundaries of urban studies. In roughly the same two hundred year period as global urbanization, human impacts on the natural environment, most notably through human-caused climate change, confront people and societies around the world with fundamental challenges to their continued ability to live where and how they wish. With a majority of humans residing in urban areas, cities are on the front lines of a range of climate change-related challenges, hazards, and, possibly, solutions, making urban questions also now climate change questions.

The nexus of urbanization and climate change challenges how cities should be understood in relation to the protection of people from the insecurities produced by environmental change. Notwithstanding the global scope of climate change, its effects are experienced at the regional level, with wide variation in the rate of warming and associated changes to weather, agriculture, animal life, and other natural systems. While climate change is happening around the world, it affects cities differently depending on their latitude, geography, topography, design, and demographics. Cities located in regions currently experiencing greater degrees of climate change or those more vulnerable to climate disruption and extreme weather events—including coastal cities, equatorial cities, and cities located at high latitudes—are already experiencing dramatic environmental changes that are disrupting ways of life and causing loss of life. One such region whose cities are in the midst of adapting to the changing climate is the circumpolar Arctic, a part of the world widely recognized as the 'canary in the coalmine' of global climate change due to its sensitivity to climate variation. Arctic regions face environmental changes and extreme weather events including, inter alia: warmer temperatures, changing precipitation patterns, ice jamming and unusual break up patterns on rivers and lakes, thawing permafrost, flooding and streamflow changes, sea-ice loss and coastal erosion, more frequent and intense storms, landslides, and wildfires. Each of these hazards can threaten the integrity and viability of urban environments, residents, and the provision of essential services. Climate change thus poses serious challenges to the integrity and viability of Arctic cities, and generates vital new questions about whether urbanization facilitates or hinders the conditions for human survival and wellbeing in a warming world.

This chapter examines the security implications of the interaction between urbanization and climate change, highlighting the links between urbanization, environmental change and human security. The primary focus of recent studies on cities and security has been on urban violence and the militarization of urban spaces through the use of military technologies for surveillance and policing (Canada 2006; Lawson et al. 2008; Graham 2010). While initially cities were not a central part of the literature on security and environmental change, even being ignored in some texts on environmental security (Floyd and Mathews 2013), that began to change after Hurricane Katrina devastated New Orleans and the U.S. Gulf of Mexico coast in 2005 (Basolo 2010; Østby et al. 2013; Simon and Leck 2013). Cities and urbanization have been included, albeit in a minor role, in environmental security research since, though often with a focus on cities in the Global South (Little and Cocklin 2010; Lemanski 2012; Romero-Lankao et al. 2013). The opposite is not the case, however; urban studies has not engaged security studies to the same degree (Davies and Imbroscio 2010). But the importance of engaging urban studies to better understand human security under conditions of environmental change is essential. Historic events such as Hurricanes Harvey and Irma in 2017 that devastated Houston, Texas and Puerto Rico, respectively, and the 2018 California wildfire season, the most destructive on record that killed 98 civilians and resulted in the destruction of entire communities, vividly demonstrate the relationship between urban environments, extreme weather events, and human security. In light of the increasing severity of extreme weather events due to climate change—and the often-overlooked degree of urbanization in the circumpolar region—the relevance of urban studies for environmental and human security in the Arctic will only grow, and demands further attention from researchers and policymakers.

In this chapter, I argue that cities are paradoxical with respect to supporting conditions of human security for their inhabitants. Urbanization represents an important development for improving global environmental sustainability by reducing people's carbon footprints and resource use. Cities simultaneously offer greater (market-based) economic opportunities for individuals to meet their basic survival needs, and can offer some resilience for people's ability to access essential goods and services and receive government support in the face of environmental disasters. However, climate change in the Arctic region complicates any assessment of the positive impacts of urbanization for human security. While cities around the world are vulnerable to disruptions caused by climate change, Arctic cities, in particular, face acute challenges as a result of changing ecological conditions. The intersection of general urban vulnerabilities to environmental change and extreme weather events, the Arctic's unique ecological sensitivity to climate change impacts, and the geographic isolation of most Arctic cities means that, as discussed in the following sections, "adapting to the impacts of climate change [...] is essential for northern communities to be secure in the decades ahead" (Canada 2009, p. 3). It remains debatable, however, whether adaptation will be sufficient for those communities and their inhabitants to remain secure as climate change accelerates over the coming decades.

I identify three reasons to doubt whether Arctic cities will be able to support and provide human security for their residents under conditions of environmental change: (1) Arctic cities are experiencing the effects of climate change in ways that undermine their critical infrastructure; (2) urbanization significantly increases local rates of warming, meaning that the growth of Arctic cities will generate a positive feedback loop that will worsen the impacts of climate change; and (3) many Arctic cities rely on fossil fuel-based industries, producing a paradox whereby they are simultaneously threatened by and reliant upon the continuation of the very economic activities responsible for contributing to global climate change. Arctic cities are thus paradoxically implicated in providing human security for their inhabitants in the face of climate change. While they may insulate people's basic wellbeing against some of the direct effects of climate change, they also expose concentrated numbers of people to the risk of critical infrastructure failures with little systems redundancy, while also deepening their reliance on industrial processes and systems that directly contribute to climate change. The current global dynamics of climate change and urbanization find concurrent expression in the Arctic region, and produce an ambivalence about whether the warming Arctic's urban future will be a bright one. This suggests that hesitation is called for towards the growth of Arctic cities that has been viewed with enthusiasm as a possible benefit of a more hospitable northern climate.

The Urban Arctic

The circumpolar Arctic region is generally understood to consist of eight states (Canada, Denmark, Finland, Iceland, Norway, Russia, Sweden, and the USA) with territory located above the Arctic Circle at a latitude of 66°33 N. Employing the definition adopted by the *Arctic Human Development Report*—which fluctuates above and below the Arctic Circle to capture specific ecological and sociological conditions—the Arctic is home to more than 4 million inhabitants, slightly fewer than half residing in Russia, and a range of unique ecosystems comprising flora and fauna that have evolved to survive in one of the most challenging environments in the world (AHDR 2004, pp. 17–18). But contrary to common depictions of the Arctic as a frozen tundra or the 'Great White North', in human terms it is a highly urbanized region. More than three quarters of the populations of Arctic countries reside in urban areas, though this is misleading as it includes the large majorities of those populations that reside outside the Arctic itself. More important is the fact that, within the Arctic regions of circumpolar states, large numbers of people reside in urban areas.

But defining 'urban' in the Arctic is challenging, as communities that would be considered too small to be meaningfully 'urban' at more southerly latitudes are often significant demographic, commercial, and administrative hubs within their northern regions, demonstrating that urbanism is more appropriately defined in relation to population and service density, rather than absolute size. Moreover, as with everything in the Arctic, there is major variation both within and between circumpolar countries. With respect to urbanization by Arctic sub-region, percentages of people living in urban areas range from a high of 94% in Iceland to a low of 33% in the Canadian territory of Nunavut (Heleniak 2014, p. 94). However, there is no standard definition of 'urban' in the Arctic, as urbanization is determined nationally according to differing criteria. This produces odd effects in the dataset of Arctic cities, for example, with Greenland considered to be 86% urbanized but Nunavut only 33% urbanized despite similar patterns of settlement in both jurisdictions.

Thus, the nature of urban spaces in the Arctic differs from those located elsewhere: "Arctic cities can be seen as an urban paradox, challenging what we know and think about what urbanity means" (Nyseth 2017, p. 59). Despite its vast geographic extent, encompassing approximately 8% of the surface of the Earth, the circumpolar region is home to a proportionately small number of people concentrated in cities and urbanized communities that are small by global standards but significant by Arctic ones. As outlined in Table 5.1, there are 13 cities in the AHDR-defined Arctic with populations greater than 50,000 people, thus meeting the US Census threshold for an urban area: Murmansk, Norilsk, Vorkuta, Apatity, Severomorsk, and Novy Urengoy in Russia; Anchorage and Fairbanks in the United States; Reykjavik, Iceland; Luleå, Sweden; Rovaniemi, Finland; and Tromsø and Bodø in Norway. Together, these 13 cities alone have a population of approximately 1.5 million people, or approximately 38% of the entire circumpolar population. This list is

Table 5.1 Arctic cities with populations greater than 50,000 people

City	*Country*	*Population (2016–2018 estimates)*[a]
Anchorage	United States	298,192
Murmansk	Russia	298,096
Norilsk	Russia	178,018
Reykjavik	Iceland	128,130
Novy Urengoy	Russia	113,254
Luleå	Sweden	76,770
Tromsø	Norway	75,638
Rovaniemi	Finland	62,667
Vorkuta	Russia	58,133
Apatity	Russia	56,356
Fairbanks	United States	51,926
Severomorsk	Russia	51,209
Bodø	Norway	51,022

[a]Measurement differences, census dates, population change, and linguistic differences make assessing the populations of Arctic cities challenging, so these figures should be taken as approximate. All population figures were accessed at www.wikipedia.org

imperfect, however, as it omits important cities usually considered 'Arctic' due to their relatively southern latitude, such as Yakutsk in eastern Russia (approximately 269,000 people), or due to their having populations of less than 50,000 people, such as Kiruna, Sweden (18,000); Nuuk, Greenland (18,000); and all three territorial capitals in northern Canada: Whitehorse (25,000), Yellowknife (19,500), and Iqaluit (8000). Defining Arctic cities exclusively in terms of population is thus problematic, as is applying any one strict definition of the Arctic region that invariably excludes some northern locales (Bennett et al. 2016).

It is difficult to make generalizations about these Arctic cities, though many share certain common characteristics. Arctic cities are typically located far apart and isolated from easy access to other regions or urban areas, resulting in cities with relatively small populations but a disproportionate degree of amenities and public services. While "Arctic cities are mostly small in scale and population, [they] represent different forms of urbanity, an urban pluralism" (Nyseth 2017, p. 60). For example, with a population of only 8000 people Iqaluit is the capital of the Canadian territory of Nunavut, represents more than 20% of the territorial population, and is its primary administrative, economic, and services hub. Nuuk is the capital city and administrative, economic, and cultural centre of Greenland, and has approximately 18,000 residents who represent nearly a third of the entire Greenlandic population. Likewise, Anchorage has fewer than 300,000 residents, but these represent more than 40% of the entire population of Alaska. And Murmansk, also with around 300,000 residents, was for many years, until recent population decline brought it below that of Anchorage, the largest city in the Arctic region and remains a major economic and military hub for northwestern Russia. In fact, Anchorage and Murmansk each represents approximately 8% of the entire circumpolar population and nearly three times more people than the entire population of northern Canada. These examples are representative of demographic and urban patterns across the region: "Along with the general trend of increased urbanization taking place across the Arctic is increased concentration of the population in most Arctic regions into one or a few of the larger urban settlements" (Heleniak 2014, p. 95). Most Arctic sub-regions have large pluralities, even majorities, of their total populations concentrated in a single urban area, such as the 76% of the Canadian territory of Yukon that resides in Whitehorse, 45% of the Northwest Territories who reside in Yellowknife, 37% of Icelanders who live in Reykjavik, and 55% of Alaskans who live in its two largest cities,

Anchorage and Fairbanks. By contrast, people in many small Arctic communities are required to travel to larger Arctic centres or cities located in the south in order to access amenities and essential services.

Despite the well-known negative consequences of climate change in the Arctic—discussed in the next section—the increasingly mild climate at higher latitudes has led some experts to predict that the northern regions of Arctic states will become increasingly habitable, and thus substantial population growth will occur, and should be encouraged, in Arctic and sub-Arctic cities (Smith 2011; Coates and Poelzer 2014; Studin 2017). Such population growth is considered essential to support Arctic economies that are reliant on natural resource extraction, tourism, and other industries that require large pools of skilled labour. Cities are already experiencing the vast bulk of demographic growth in the Arctic, whose population has shrunk slightly since 2000 but is expected to grow by around 4% by 2030 (Heleniak 2014, p. 100). However, as ever, the aggregate regional data masks substantial variation, since healthy rates of growth in the North American and European Arctic regions will be balanced by continued steep decline in northern Russia that result in a modest average rate of population growth across the region. Despite having the largest absolute population in the Arctic, northern Russia has experienced dramatic decline since the collapse of the Soviet Union, and "the predominant trend across the Russia[n] Arctic is absolute population decline in the oblast centers or largest settlements combined with increases in their shares of the regions' overall population. This obviously indicates significant depopulation of the areas outside of the largest settlements, including in many cases, complete closure of many smaller settlements" (Heleniak 2014, p. 97). By contrast, population growth in northern Canada and northern Norway since 2000 has been as high as 10–20%. The growth of larger Arctic cities has also occurred at the expense of smaller communities, which continue to shrink and some of which have been or may be abandoned entirely due to the combined effects of environmental change and the high costs of living and of providing government services in far-flung small communities that are often only accessible for part of the year by water, sea-plane, or ice road.

The central point is that notwithstanding its vast land area and the presence of many small communities—particularly those that are now home to Indigenous peoples who have lived on the land for centuries or millennia—the Arctic is a highly urbanized region that can be appropriately examined from the perspective of urban studies.

However, the specificities of the Arctic and its unique ecological context also make the application of urban models and expectations problematic, since these fail to take into account the distinct needs and challenges experienced by Arctic cities. Foremost in this respect are the effects of climate change on Arctic urban areas, and the ways in which Arctic climate change undermines the expectation that the application of technology and sustainable urban planning may allow cities to become a bulwark against the insecurities produced by climate change. The nature of Arctic climate change, and its effects on Arctic cities, is examined in the following section.

The Changing Arctic Environment

Earth's climate is changing due to human interference in the global climate system. Greenhouse gas (GHG) emissions primarily resulting from the burning of fossil fuels, deforestation, and other land use changes are causing increased retention of solar energy that is warming the planet's surface. The Intergovernmental Panel on Climate Change (IPCC) indicates that average global temperature has increased by nearly 1 °C since 1880, causing species extinction, biodiversity loss, desertification, rising sea levels, coastal flooding, declining crop yields, reduced freshwater, increased frequency and severity of extreme weather events, and displacement of vulnerable people, particularly in the Global South (IPCC 2013; World Bank 2013). Increasing concentrations of GHGs will force climate change of a further 2–3 °C throughout the twenty-first century, and "caus[ing] further warming and changes in all components of the climate system" unless mitigation occurs (IPCC 2013, p. 19).

Sufficient mitigation is, however, unlikely. The World Bank estimates "recent emissions and future emission trends imply higher 21st century emissions levels than previously projected. As a consequence, the likelihood of 4 °C warming being reached or exceeded this century has increased [...] In the absence of further mitigation action there is a 40% chance of warming exceeding 4 °C by 2100 and a 10% chance of it exceeding 5 °C in the same period" (World Bank 2013, p. xv). Combined annual rates of natural and human-caused emissions exceed any period in the last 22,000 years, and atmospheric GHG concentrations are higher than any point in the past 800,000 years, with the planet on track to exceed the worst-case scenarios modelled by the IPCC (2013, p. 11). These conditions are well outside the range in which human civilization has developed, and will result in a radically different global climate from that in which current social, demographic, and geographic configurations emerged. Moreover, recent data suggest climate models are conservative in their estimates, and acute environmental changes are occurring faster than predicted (Derksen and Brown 2012; Scherer 2012). Increasing global temperature has raised concern over the prospect of 'tipping points' that might catalyse rapid, uncontrollable threshold effects with catastrophic consequences for ecological integrity and resilience around the globe (Lenton et al. 2008; Lenton 2011). In particular, scientists worry positive feedbacks generated by climate change may exacerbate global warming, including, inter alia, release of methane from thawing Arctic and sub-Arctic permafrost, reduced reflectivity of solar radiation (albedo) due to diminished polar ice, or the collapse of the Greenland or Antarctic ice-sheets (IPCC 2013, p. 16).

The concentration of these feedbacks in the polar regions is not coincidental. As the most sensitive areas to climate variation, the poles are the proverbial 'canaries in the coalmine' experiencing the fastest and most dramatic climate change on the planet. But whereas Antarctica has a less varied climate and no permanent inhabitants, warming temperatures are reshaping Arctic ecologies and societies in profound ways. Numerous studies, notably the *Arctic Climate Impact Assessment* and the IPCC's *Fourth* and *Fifth Assessment Reports*, document the changes occurring in the Arctic. Sea ice declined by 9–13% per decade between 1979 and 2012, reaching an historic low nearly 50% below the average 1979–2000 extent in the summer of 2012 (IPCC 2013, p. 9; Larsen et al. 2014b, p. 3). Warming is occurring faster than predicted, and some expect the Arctic to be free of summer sea ice in as few as 30 years, decades before it was predicted (Wang and Overland 2009). Climate records continue to be broken, and dramatic changes include more extreme seasonal variation, reduced sea ice, receding glaciers, diminished snow cover, thawing permafrost, changing terrestrial water systems, invasive species, temperatures increasing at twice the global average, and other stressors on plant and animal populations.

The scope and scale of this climate transformation is such that it is altering the ecological basis upon which all past human activity in the Arctic has been based. Climate change is affecting, and undermining, established ways of living in one of the harshest and most challenging survival environments in the world. As discussed in the next section, many of these changes can be understood as threatening human security. According to the *Arctic Climate Impact Assessment*: "The sum of these factors threatens to overwhelm the adaptive capacity of some Arctic populations and ecosystems. The increasingly rapid rate of recent climate change poses new challenges to the resilience of Arctic life" (ACIA 2004, p. 5). This view is confirmed by the IPCC's *Fifth Assessment Report*, which notes: "The rapid rate at which climate is changing in the Polar Regions will impact natural and social systems and may exceed the rate at which some of their components can successfully adapt" (Larsen et al. 2014b, p. 3). Thus, the Arctic faces unprecedented changes and challenges across two distinct axes: the increasing urbanization of its inhabitants, and the radical and rapid transformation of the ecosystem on which all aspects of life in the Arctic has been based.

Arctic Cities and Human Security

As a result of these impacts, environmental change in the Arctic has become widely understood as producing threats to human security for people across the region (Dalby 2009; Hoogensen Gjørv et al. 2014; Greaves 2016a). The concept of human security—often attributed to the United Nations Development Program's (UNDP) 1994 *Human Development Report*, though with earlier roots—seeks to change the ontology of security from one primarily committed to the survival and interests of states to one concerned for the survival and wellbeing of people (UNDP 1994). It explicitly broadens security analysis to include potential insecurities other than armed force and military threat, and to deepen its perspective to include referent objects below the level of the sovereign state. The view of security as necessarily both broad and deep forms the crux of the UNDP approach, subsequently adopted within the UN system and by various other political actors, which defined human security as "safety from such chronic threats as hunger, disease, and repression [...] and protection from sudden and hurtful disruptions in the patterns of daily life" (UNDP 1994, p. 23). Capturing the need for people to have 'freedom from fear' (violence and repression) and 'freedom from want' (chronic suffering and privation) in order to be considered secure, the report illustrates the variety of potential harms by specifying seven dimensions of human security: economic, food, health, environmental, personal, community, and political. Although its applicability in the Arctic has been debated (Griffiths 2008; Hoogensen Gjørv 2014), I have argued elsewhere that human security provides an analytically useful framework for understanding the nature of security threats in the circumpolar Arctic, particularly with respect to the effects of human-caused environmental change (Greaves 2012a, b). Environmental changes have profound implications for medium and long-term human security in the Arctic, and require significant efforts to facilitate adaptation and mitigation of their worst impacts.

Environmental hazards in the Arctic such as climate change, transboundary pollution, ecological degradation, and resource depletion are human security issues, namely issues with significant negative impacts on human wellbeing, and which, in extremis, undermine prospects for human survival. Researchers have identified up to 11 direct and indirect climate related impacts on Arctic human health, including: new physical hazards related to the changing landscape, increased rates of accidents and fatalities due to unpredictable ice and weather patterns, new vectors for communicable disease, changes to food- and water-borne pathogens, increased exposure to environmental contaminants, and ozone depletion causing increased exposure to ultraviolet radiation (Hild and Stordahl 2004; Séguin 2008). Climate change worsens existing transboundary pollution such as mercury and persistent organic pollutants (POPs), which increase rates of cancer and cause neurological damage in children, and are aggravated by pollutants such as black carbon and contamination from military and industrial activity conducted during the Cold War (Downie and Fennge 2003; AMAP 2011). POPs are particularly dangerous because they concentrate within individual organisms and bio-magnify as they concentrate at increasingly high levels of the food chain, culminating in humans. POP contamination is directly worsened by climate change because warmer temperatures increase their toxicity and speed their rate of degradation into the environment (Noyes et al. 2009).

This pollution has raised concerns over nutrition and food security as the quality and availability of traditional food sources has declined, particularly large mammals, fish stocks, and plant life (Hild and Stordahl 2004; Meakin and Kurvits 2009). This, in turn, exposes Arctic populations to greater vulnerability as local economies transition towards market-based waged systems and people and communities become increasingly unable to support themselves using traditional subsistence methods (Duhaime 2004; Aarsæther et al. 2004). It also affects Indigenous cultures and spirituality that are closely linked to the natural environment, including animal populations whose hunting and consumption is central to cultural and spiritual practices (Csonka and Schweitzer 2004). Environmental changes thus exacerbate existing challenges to human security already present in the Arctic, such as poverty, severe climate, vulnerability to extreme weather events, and the socio-cultural legacies of colonialism and forced assimilation into dominant southern cultures.

However, despite widespread concern over the implications of climate change for the regional ecology and its human inhabitants, particularly Indigenous peoples, Arctic states have not constructed climate change or its direct effects as threats to security in the region. Instead, state understandings of Arctic security emphasize traditional and unconventional military threats to their territorial sovereignty, and the extraction of natural resources considered essential to the interests of Arctic states, and to the global economy (Heininen 2012; Greaves 2016b). Though climate change is the catalyst for these re-imaginings of Arctic security, circumpolar states have not considered the impacts of climate change as the principal hazards to be secured against. Rather, Arctic security is understood to relate to protection against second order phenomena enabled by climate change, or, paradoxically, as pursuit of economic activity directly linked to furthering climate change. This implicitly views the increase in Arctic populations that are necessary to support increased economic activity in the northern regions of circumpolar states as contributing positively to security in the region.

The Arctic has long been considered exceptional, essentially thought to be "a unique region detached, and encapsulated, from global political dynamics, and thus characterized primarily as an apolitical space of regional governance, functional cooperation, and peaceful co-existence" (Käpylä and Mikkola 2015, p. 4). However, the effects of climate change and globalization have combined to increasingly integrate the Arctic into global political and economic competition, reducing the peaceful distinctiveness that had previously set the Arctic apart. While still peripheral, the Arctic is now widely understood to be implicated in processes and institutions that transcend the circumpolar region itself. Arctic life is fundamentally shaped by forces originating far outside the region itself, particularly global market demand for commodities, non-Arctic politics, and, indeed, demographic factors such as population migration. The Arctic of the twenty-first century is part of a globalized world, and reflects global trends, such as increased urbanization. However, the distinct features of Arctic cities suggest that the region remains exceptional in various ways, such as with respect to the implications of urbanization and environmental change for human security.

In their book, *Resilient Cities*, Newman et al. (2009) outline four scenarios for the future of cities in a world defined by climate change and high energy costs due to peak oil. Three scenarios produce problematic outcomes for sustainability and the provision of security and well-being for urban inhabitants: cities may collapse due to their fossil-fuel intensive design and lack of public infrastructure; they may become ruralized as low-density suburban communities enable small-scale agriculture and distributed energy production in people's homes and backyards; or they may become divided between impoverished, inefficient suburban and exurban regions and high-density, high-efficiency urban cores with quality public services such as public transit enabling urban mobility and access to employment. The authors' preferred scenario is that cities will become resilient, wherein "every step of development and redevelopment of the city will make it more sustainable: it will reduce its ecological footprint […] while simultaneously improving its quality of life (environment, health, housing, employment, community) so that it can better fit within the capacities of local, regional, and global ecosystems" (Newman et al. 2009, p. 7). Indeed, 'resilience' is a favoured lens through which to view the adaptation needs of cities in the circumpolar region, as reflected in the establishment of the Arctic Resilient Cities Network under the auspices of the United Nations Human Settlements Programme (UN-Habitat) and the work of the Arctic Mayor Forum. Most literature on urban resilience under conditions of global climate change emphasizes that reliance upon fossil fuel energy is the kiss of death for future urbanization; cities that require fossil fuels to operate will place increasing burdens upon their residents, while simultaneously failing to adapt to the decarbonizing future required to avoid the worst predicted effects of human-caused environmental change. In effect, fossil fuel reliance is incompatible with urban resilience.

The ideal of resilience raises serious questions for Arctic cities, which are not only heavily reliant on fossil fuels for the provision of essential goods, but also on fossil fuel-based economies for employment and tax revenues. As such, there are at least three reasons to question the vision of an increasingly urbanized Arctic as a model for the region's future, particularly insofar as human security in the Arctic is considered to be promoted through greater urbanization. The assumption that in the context of a changing environment human security—the wellbeing of people and the conditions of their survival and prosperity—is more likely or can be better assured in cities, may be an overly optimistic assessment of the region's future.

First, and most immediately, the warming environment is undermining the physical integrity of vulnerable critical infrastructure necessary to support Arctic cities. Five categories of infrastructure are particularly relevant: transportation, buildings, communications, energy, and storage and containment facilities for mine waste and tailings. As described by the Government of Canada, "maintaining and enhancing access to reliable infrastructure that provides mobility, shelter, connectivity, power, and protection from toxic industrial waste is at the core of sustainable regional development and northern security" (Canada 2009, p. 48). Infrastructure built on permafrost or ice—such as ice-roads and bridges—are exceptionally vulnerable to permafrost thawing and warmer temperatures. Impacts are already apparent in damage to critical infrastructure such as roads, highways, bridges, airstrips, pipelines, sewage systems, homes, and other buildings as a result of the destabilization of the ground upon which many Arctic communities are built (ACIA 2004; Auld et al. 2007). In some Siberian communities, including large cities such as Norilsk and Vorkuta, between 40 and 60% of buildings have already been damaged by thawing permafrost beneath their foundations (Schreiber 2018). While damage to the first four types of infrastructure is serious enough, the prospect of a tailings pond or mining waste rock facility being compromised by thawing permafrost resulting in seepage or collapse is particularly alarming, and some analysts consider it to be an underappreciated possibility in many parts of the Arctic (Sohns 2017). Such industrial accidents have occurred elsewhere, with catastrophic consequences for plant and animal populations, and freshwater supplies for communities and ecosystems located downstream. In the case of any critical infrastructure failures, however, the implications for human security are compounded by the likelihood that there is limited redundancy or duplication in most Arctic communities, even in larger cities. If compromised, critical infrastructure necessary to support human security will take considerable time and resources to repair.

Many of these hazards are exacerbated by rising sea levels and coastal erosion. Estimated rates of glacier melt are now expected to raise global sea levels by as much as 10 feet over the next century (Gillis and Chang 2014), and "in some cases, communities and industrial facilities in coastal zones are already threatened or being forced to relocate" (Durbak and Strauss 2005, p. 134). Some smaller communities in Alaska have been relocated already as coastal erosion renders them uninhabitable (ACIA 2004, p. 80; DeMarban 2014), with as many as 178 villages identified by the US Army Corps of Engineers as threatened (EPA, n.d.). Similar fears exist for communities in Canada, Greenland, and Russia (ACIA 2004, p. 81). While larger cities have not yet been as severely affected, in the future Arctic cities located on coastlines (which is most of them), will also experience sea level rise and higher likelihood of extreme weather events, including in places that have not historically experienced them, and are thus poorly prepared. The possibility of an alteration in the Gulfstream, the current of warm water from the south Atlantic Ocean to the north Atlantic Ocean that accounts for the relatively moderate climates of Iceland, Scandinavia, and the eastern Canadian Arctic, is particularly worrisome in this regard, and may paradoxically result in the cooling of these regions as well as a higher likelihood of extreme weather events such as storms and hurricanes. Some of these effects can be mitigated through smart design choices that account for both current Arctic conditions and how these are expected to change in future. However, best practices associated with urban design also have to be updated to reflect climate change, and are not currently equipped to address the scale of these challenges: "The urban design principles for winter settlements may need to be widened in scope to encompass new winter conditions associated with climate change [...] However, because climate chance will affect different winter settlements in different ways, designers should seek to understand its impact in the local context, and avoid generic 'one size fits all' approach to climate mitigation, adaptation or resilience" (Chapman et al. 2018, p. 15). In effect, climate impacts make Arctic cities with advanced infrastructure more vulnerable to disruption by climate change, as well as more expensive to build and maintain, and more difficult to adapt or modify to reflect standards or principles applied elsewhere. They also pose direct and indirect threats to human health and survival.

Geographic isolation and heavy reliance on imported goods make daily life in many Arctic cities inherently vulnerable to climate disruption, which can thus endanger human security in ways that do not apply to urban areas located at more southerly latitudes. Smaller Arctic communities are even more vulnerable to such disruption, such as the 84 communities located across the Canadian North that lack year round surface transportation, resulting in extremely high prices for foods and consumer goods. The number of communities in northern Canada accessible by road is extremely limited, with most communities relying on local airports and small harbours for access to the outside world. The same is true of Greenland, where no two towns are connected by road and where most communities exist in a precarious string along the country's western and southern coasts, serviced by small airports and a single ferry boat. However, larger urban centres that play key functions within their regions are not immune to this vulnerability. Yellowknife, Northwest Territories, for instance, is accessible only by a single highway that can be disrupted by seasonal weather, and is also threatened by thawing permafrost. Alaska Highway 2 is the only road access east of Fairbanks connecting the state overland with Yukon, in Canada, and southwards towards the continental United States. As another example, the ports at Nuuk, Tromsø, and Iqaluit are vital for supplying food and other goods within their respective Arctic regions, and any significant damage or disruption—for example, due to a severe storm, hurricane, or coastal erosion—would severely impact the capacity of nearby populations to receive essential supplies. Whereas urban services and systems in other regions of the world may help insulate people's human security from climate change by allowing access to essential goods and services, including disaster and emergency response, many Arctic cities rely on precarious supply chains for resources such as food and energy, particularly diesel fuel needed to power electricity generators.

In addition, the question of urbanization is also implicated in the human security of Arctic Indigenous peoples, since Indigenous ontologies of wellbeing are premised upon relationship to the land that is often undermined when individuals move to urban areas. Indigenous human security is based on an ontological relationship to land and maintenance of Indigenous cultural practices, which are negatively affected by both urbanization and human-caused environmental changes to the land itself, both through climate change and extractive activities (Greaves 2016b). Urbanization, furthermore, has mixed impacts on indicators of Indigenous health and wellbeing; levels of substance abuse, addiction, inadequate housing, hunger and food insecurity, domestic violence, and incarceration typically remain high relative to non-Indigenous populations, but individuals are better able to access health and social services, educational opportunities, and employment (Ellsworth and O'Keeffee 2013; Rautio et al. 2014; Smylie and Firestone 2017). Indigenous peoples remain a minority in Arctic cities outside of Greenland and Nunavut, and there are considerable methodological challenges in gathering and assessing data on urban Indigenous health. However, studies of circumpolar health have identified "living close to nature" or similar variables as an important determinant of health in the region (Rautio et al. 2014, p. 322), suggesting that time spent on the land and the exercise of land-based cultural practices are important for Indigenous health, and that conditions in urban environments may be detrimental for health in this way. The irony is considerable, given that it was the governments of Arctic states that enacted policies during the twentieth century designed to encourage the sedentization and permanent settlement of Indigenous peoples. These policies had the intended effect while also disrupting Indigenous lifeways in fundamental and deleterious ways. They also have reduced the resilience of Indigenous peoples by increasing their reliance on government and market-based economic activities for the provision of essential goods and services. Permanent settlement, even in small communities, has thus produced somewhat ambivalent conditions of human security for Arctic Indigenous peoples under environmental change.

The second reason to question urbanization and human security under global climate change is that, while the effects of climate change on cities are widely recognized, the impacts of cities on climate change have been underappreciated by the field of climate science and climate policy. However, as demonstrated by Brian Stone (2012) and colleagues at the Urban Climate Institute at the Georgia Institute of Technology, urbanization has a pronounced impact on increasing the local impacts of climate change. They show that changes in land use (mostly the conversion of natural systems to built environments), combined with the concentration of infrastructure and human bodies creates an urban heat island effect that can produce local climate change that is as much as 10× greater than background levels of global warming. Put another way, while the world has warmed by around 1 °C from the pre-industrial average, the increase in urban temperatures ranges from 1.5 to 11 °C higher than non-urban surrounding areas (Stone 2012, p. 77). While urban warming in Artic cities is likely at the lower end of this range due to their high latitudes and relatively modest size, higher rates of local warming will still contribute to the damage to critical infrastructure noted above, while weakening local ecosystems' adaptive capacities, particularly with respect to freshwater shortages in and around urban areas.

Vivid examples of this warming—both the global warming trend and urban warming—have been evident in Arctic cities in recent years, particularly exceptionally warm summer temperatures in high latitude cities in Siberia and the central and western Canadian Arctic, such as the summer of 2015 when Whitehorse, Yukon—at nearly 61° latitude—had the highest average summer temperature of any city in Canada (CBC News 2015). In this way, the growth of Arctic cities will increase the impacts of climate change on those urban areas, generating a positive feedback loop that will worsen the local impacts of climate change. On top of this, the fact is that people in Arctic cities already have high GHG emissions per capita due to energy for heating in winter, emissions from flying (especially fly-in communities), and reliance on emission intensive energy such as diesel. Increasing the populations of Arctic cities will increase the levels of emissions associated with these populations while simultaneously increasing local rates of warming due to land use changes as more area is converted from natural to urban environments.

The third reason is that many Arctic cities rest uneasily on a paradox derived from their current reliance on fossil fuel-based or -intensive extractive industrial sectors for employment, tax revenues, and investment, as well as expected future economic growth (Huskey et al. 2014). As a result of a US Geological Survey report estimating the Arctic contains 90 billion barrels of oil (13% of undiscovered global resources) and 46 trillion cubic metres of natural gas (30% of undiscovered global resources), since 2009 there has been intense international interest in the expanded extraction of Arctic resources to serve global energy demand (USGS 2008). Driven by the boom in global commodity prices prior to 2014, the Arctic was considered to be the next frontier for global fossil fuels, and, conversely, the energy sector was seen to be the answer to the questions of how to modernize many Arctic economies and have circumpolar regions contribute greater wealth to their respective national economies. However, the anticipated Arctic energy boom may prove illusory. The collapse in the global price of oil from its high of over US$140 per barrel to around $40 per barrel between 2014 and 2019 has largely cooled private sector enthusiasm for new conventional Arctic energy resources. While Arctic production has continued and even expanded in parts of Russia (Gasper 2018), major energy companies have cancelled or suspended their Arctic projects (Eilperin and Mufson 2015). Major re-insurers have also indicated their growing unwillingness to finance Arctic oil and gas drilling, flagging concerns over liability related to both local environmental impacts and possible litigation related to climate change (Green 2019). Changes to public policy in Arctic states and elsewhere around the world designed to curb GHG emissions and reduce fossil fuel consumption, such as Canada's federally imposed carbon tax, will likely further depress both private sector interest in Arctic energy investment and public demand for hydrocarbons. Overall, while large scale resource extraction remains one of the three central pillars of Arctic economies, the *Arctic Human Development Report* may have been correct to note, in 2014, that "together these effects suggest that high and rising prices of Arctic resources may not be a long-term phenomenon [...] Arctic resources may face the long run pattern of resource prices declining in real terms" (Huskey et al. 2014, p. 161). This decline in investment in Arctic resources will pose real challenges to the political economies of many regions across the Arctic.

The fact that the very same human-caused climate change that is enabling greater access to Arctic hydrocarbon resources will be exacerbated and accelerated by the extraction and consumption of those resources is, from an urban studies perspective, compounded by the fact that Arctic cities require continued expansion of these industrial activities to maintain their current economic and social model.[1] Therein lies the challenge: the extraction and burning of fossil fuels in the Arctic is but a small part of the global context of GHG emissions and land use changes contributing to climate change. Ending such activities in the Arctic, even were it politically feasible, would not halt global warming or 'save' the Arctic from its warmer future. But the success of global efforts to combat climate change, as reflected in the multilateral Paris Agreement commitment to limit warming to less than 2 °C that has been signed by 194 states since 2015, requires that some fossil fuels remain undeveloped. Given the higher cost of developing Arctic hydrocarbons, one scientific assessment is "that all Arctic [energy] resources should be classified as unburnable," and undeveloped Arctic reserves must remain unexploited if the planet is to avoid the catastrophic effects of warming by more than 2 °C (McGlade and Ekins 2015, p. 190). The result of this paradox is that for the economies of many Arctic cities to flourish according to current plans and business models, the international community must fail in its efforts to curb global warming and dramatically reduce the emissions of GHGs into the atmosphere.

But even if Arctic resource extraction goes forward under a 'business as usual' scenario, with attendant benefits for circumpolar economies, the Arctic environment will be fundamentally altered by the effects of climate change such that impacts upon communities, peoples, and entire regions will be devastating. Former US President Barack Obama (2015) described the predicted effects of continued GHG emissions at length in a keynote address to a climate change conference in Anchorage:

> If those trend lines continue the way they are, there's not going to be a nation on this Earth that's not impacted negatively. People will suffer. Economies will suffer. Entire nations will find themselves under severe, severe problems [...] Submerged countries. Abandoned cities. Fields no longer growing. Indigenous peoples who can't carry out traditions that stretch back millennia. Entire industries of people who can't practice their livelihoods. Desperate refugees seeking the sanctuary of nations not their own. Political disruptions that could trigger multiple conflicts around the globe [...] On this issue, of all issues, there is such a thing as being too late. That moment is almost upon us.

The seriousness of unmitigated GHG emissions were reiterated by a 2018 report of the IPCC outlining pathways to keep global warming to the relatively safe threshold of 1.5 °C (IPCC 2018).

The truth is that there may be no easy future for many cities across the circumpolar region. Without, or with smaller and declining fossil fuel sectors, what will be the economic base that provides employment to support the growing Arctic populations in Alaska, Norway, Greenland, and Nunavut? But even worse, if the fossil fuel sector does grow, if we fail in our efforts to curb the fundamental cause of climate change rooted in human consumption of fossil fuels, what kind of life will Arctic cities even be able to support? In the latter scenarios, the damaging effects of climate change outlined above will be even worse, and the adaptive capacities of Arctic cities even more challenged to provide security for their 1.5 million inhabitants.

Conclusion: Northern Futures?

There is an ambivalent relationship, at best, between urbanization and human security across a rapidly warming circumpolar Arctic, and it is likely that the negative effects of climate change for the resilience of human life in Arctic cities has been underestimated. While there has been distinct enthusiasm in many quarters around population growth, urbanization, and increasing marketization of the Arctic, I have outlined here at least three reasons to doubt both the likelihood and desirability of unfettered growth of Arctic cities. Based on the analysis in this chapter, projections of Arctic urbanization and population growth are based on fundamentally status quo theorizing or 'business as usual projections' that: (a) downplay significant policy shifts to price carbon and increase the cost of business; (b) expect continued high global demand for fossil fuels and other Arctic resources; and (c) exclude the possibility of extreme climate events or catastrophic climate change, which would clearly be negative for human security. Practically speaking, there is only room for things to get worse in terms of Arctic sustainability and conditions for supporting human security. Existing Arctic cities are not going anywhere anytime soon (Arctic cities will, and should, continue to exist), but the rate of growth and policy support for increased urbanization should be contingent on further assessment of the viability of Arctic urbanization.

By contrast, there may be a middle ground for the promotion of human security in the Arctic by pursuing policies that support the maintenance of mid-size and smaller Arctic communities. Smaller urban areas that occupy a smaller geographic footprint while maintaining population densities capable of supporting services and economic activities and reducing the need for vehicular transportation will help mitigate the exacerbating effects of urbanization on local climate warming. Smaller urban areas will put less pressure on their regional ecosystems, and thus allow for a greater degree of local sustenance (particularly country foods). Human security in the Arctic, in terms of food and water security and maintaining strong local communities, will help to be sustained by supporting the distinctive composition of market and subsistence-based economies in much of the Arctic. This will also help support human security for Indigenous peoples, since smaller communities allow for easier transmission of land-based knowledge, skills, and cultural practices, rather than encouraging relocation to distant, albeit still Arctic, urban centres that remove people from their connection to the land. Crucially, the benefits of smaller urban communities for human security are contingent upon access to essential services such as health care being adequately funded and provided. However, overall, resilience and wellbeing will be promoted by limiting (though not eliminating) community reliance on high cost imported goods, particularly food, through global supply chains that will be made increasingly tenuous and expensive in a climate and resource stressed future.

The future will be warmer, less stable, and more challenging for peoples living all over the globe, particularly in vulnerable ecosystems. While there are benefits for the security and wellbeing of people that can be achieved through increased urbanization, these are variable and contingent on local ecological and socio-economic factors. The Arctic, long considered a distinct or exceptional region of the world, will become increasingly like the rest of the planet as a result of climate change. However, conditions that support human security in the Arctic will remain distinct from those elsewhere due to the specific factors that will continue to make it a region of small populations, clustered in isolated urban environments, peripheral to the global economy and global political institutions. While it is possible that a significant degree of human security can be attained for Arctic residents in the coming decades as their region transforms, this will not occur by adopting the same policy prescriptions as the rest of the world. While perhaps less exceptional than in its mythic past, the Arctic region, and its cities, remain distinct, and their circumstances in a warming world will differ from those elsewhere. For people in the Arctic, security will not come entirely from cities, but from finding a balance between the provision of urban services and maintaining ties to the land which continues to be central to life across the North.

Note

1. For a discussion of the pathologies of contemporary approaches to Arctic security, climate change, and hydrocarbon extraction see Andrew Chater and Wilfrid Greaves. 2014. "Security Governance in the Arctic," in *Handbook on Governance and Security*, ed. Jim Sperling, pp. 123–147. Northampton: Edward Elgar.

References

Aarsæther, Nils, Larissa Riabova, and Jørgen Ole Bærenholdt. 2004. Community Viability. In *AHDR, Arctic Human Development Report*, 139–154. Akureyri: Steffanson Arctic Institute.

ACIA. 2004. *Impacts of a Warming Climate: Arctic Climate Impact Assessment*. Cambridge: Cambridge University Press.

AHDR. 2004. *Arctic Human Development Report*. Akureyri: Steffanson Arctic Institute.

AMAP. 2011. *Snow, Water, Ice and Permafrost in the Arctic (SWIPA): Climate Change and the Cryosphere*. Oslo: Arctic Monitoring and Assessment Program.

Auld, Heather, Don MacIver, and Joan Klaassen. 2007. *Adaptation Options for Infrastructure Under Changing Climate Conditions*. Ottawa: Environment Canada. www.hazards.ca. Accessed 18 Jan 2016.

Basolo, Victoria. 2010. Environmental Change, Disasters, and Vulnerability: The Case of Hurricane Katrina and New Orleans. In *Global Environmental*

Change and Human Security, ed. Richard A. Matthew, Jon Barnett, Bryan McDonald, and Karen L. O'Brien, 97–116. Cambridge: The MIT Press.

Bennett, Mia, Wilfrid Greaves, Rudolf Riedlsperger, and Alberic Botella. 2016. Articulating the Arctic: Contrasting State and Inuit Maps of the Canadian North. *Polar Record* 52 (6): 630–644.

Canada. 2006. *Human Security for an Urban Century: Local Challenges, Global Perspectives*. Ottawa: Canadian Consortium on Human Security.

Canada. 2009. *True North: Adapting Infrastructure to Climate Change in Northern Canada*. Ottawa: National Roundtable on the Environment and the Economy.

CBC News. 2015. Yukon's Record Breaking Temperatures Enter Second Week. May 24. http://www.cbc.ca/news/canada/north/yukon-s-record-breaking-temperatures-enter-second-week-1.3085654. Accessed 9 Jan 2019.

Chapman, David, Kristina L. Nilsson, Agatina Rizzo, and Agneta Larsson. 2018. Updating Winter: The Importance of Climate-Sensitive Urban Design for Winter Settlements. In *Arctic Yearbook 2018*, 1–20. Accessed from https://arcticyearbook.com/images/yearbook/2018/Scholarly_Papers/5_AY2018_Chapman.pdf.

Chater, Andrew, and Wilfrid Greaves. 2014. Security Governance in the Arctic. In *Handbook on Governance and Security*, ed. Jim Sperling, 123–147. Northampton: Edward Elgar.

Coates, Ken, and Greg Poelzer. 2014. The Next Northern Challenge: The Reality of the Provincial North. In *MLI Commentary*. Ottawa: MacDonald-Laurier Institute.

Csonka, Yvon, and Peter Schweitzer. 2004. Societies and Cultures: Change and Persistence. In *AHDR, Arctic Human Development Report*, 45–68. Akureyri: Steffanson Arctic Institute.

Dalby, Simon. 2009. *Security and Environmental Change*. Cambridge: Polity.

Davies, Jonathan S., and David L. Imbroscio (eds.). 2010. *Critical Urban Studies: New Directions*. Albany: SUNY Press.

DeMarban, Alex. 2014. Eroding Alaska Village Urges Congress to Address Climate Change. *Alaska Dispatch*, January 16. http://www.alaskadispatch.com/article/20140116/eroding-alaska-village-urges-congress-address-climate-change. Accessed 8 Jan 2019.

Derksen, Chris, and Ross Brown. 2012. Spring Snow Cover Extent Reductions in the 2008–2012 Period Exceeding Climate Model Projections. *Geophysical Research Letters* 39 (19): 1–6.

Downie, David L., and Terry Fenge (eds.). 2003. *Northern Lights Against POPs: Combatting Toxic Threats in the Arctic*. Montreal-Kingston: McGill-Queen's University Press.

Duhaime, Gérard. 2004. Economic Systems. In *AHDR, Arctic Human Development Report*, 69–84. Akureyri: Steffanson Arctic Institute.

Durbak, Christine K., and Claudia M. Strauss. 2005. Securing a Healthier World. In *Human and Environmental Security: An Agenda for Change*, ed. Felix Dodds and Tim Pippard, 128–138. London: Earthscan.

Eilperin, Juliet, and Steven Mufson. 2015. Royal Dutch Shell Suspends Arctic Drilling Indefinitely. *The Washington Post*, September 28. https://www.washingtonpost.com/news/energy-environment/wp/2015/09/28/royal-dutch-shell-suspends-arctic-drilling-indefinitely/?utm_term=.aa299882a45e. Accessed 9 Jan 2019.

Ellsworth, Leanne, and Annmaree O'Keeffee. 2013. Circumpolar Inuit Health Systems. *International Journal of Circumpolar Health* 72 (1): 1–9.

EPA. n.d. Coastal Alaska Native Villages Plan for Relocation. Adaptation Examples in Alaska. http://www.epa.gov/climatechange/impacts-adaptation/alaska-adaptation.html. Accessed 8 Jan 2019.

Floyd, Rita, and Richard A. Matthew (eds.). 2013. *Environmental Security: Approaches and Issues*. New York: Routledge.

Gasper, Donald. 2018. China and Russia Want to Develop Arctic Energy Resources Together, and US Disapproval May Not Deter Them. *South China Morning Post*, September 12. https://www.scmp.com/comment/insight-opinion/asia/article/2163719/china-and-russia-want-develop-arctic-energy-resources. Accessed 4 Feb 2019.

Gillis, Justin, and Kenneth Chang. 2014. Scientists Warn of Rising Oceans from Polar Melt. *The New York Times*, May 12. http://www.nytimes.com/2014/05/13/science/earth/collapse-of-parts-of-west-antarctica-ice-sheet-has-begun-scientists-say.html?_r=0. Accessed 15 May 2014.

Graham, Stephen. 2010. *Cities Under Siege: The New Military Urbanism*. London: Verso.

Greaves, Wilfrid. 2012a. For Whom, From What? Canada's Arctic Policy and the Narrowing of Human Security. *International Journal* 67 (1): 219–240.

Greaves, Wilfrid. 2012b. Insecurities of Non-dominance: Re-theorizing Human Security and Environmental Change in Developed States. In *Natural Resources and Social Conflict: Towards Critical Environmental Security*, ed. Matthew A. Schnurr and Larry A. Swatuk, 63–82. New York: Palgrave.

Greaves, Wilfrid. 2016a. Securing Sustainability: The Case for Critical Environmental Security in the Arctic. *Polar Record* 52 (6): 660–671.

Greaves, Wilfrid. 2016b. Arctic In/Security and Indigenous Peoples: Comparing Inuit in Canada and Sámi in Norway. *Security Dialogue* 47 (6): 461–480.

Green, Miranda. 2019. Barclays to Reject Most Financing for Arctic Drilling. *The Hill*, January 14. https://thehill.com/policy/energy-environment/425252-barclays-to-reject-most-financing-for-drilling-in-arctic. Accessed 9 Jan 2019.

Griffiths, Franklyn. 2008. Not That Good a Fit? 'Human Security' and the Arctic. Arctic Security in the 21st Century. In *Conference Report Co-convened by the Simons Foundation and the School for International Studies, Simon Fraser University*. Vancouver: Simon Fraser University.

Heininen, Lassi. 2012. State of the Arctic Strategies and Policies—A Summary. In *Arctic Yearbook 2012*, 2–47. Akureyri: Northern Research Forum.

Heleniak, Timothy. 2014. Arctic Populations and Migration. In *Arctic Human Development Report: Regional Processes and Global Linkages*, ed. Joan Nymand Larsen and Gail Fondahl, 53–104. Akureyri: Steffanson Arctic Institute.

Hild, Carl M., and Vigdis Stordahl. 2004. Human Health and Well-Being. In *AHDR, Arctic Human Development Report*, 155–168. Akureyri: Steffanson Arctic Institute.

Hoogensen Gjørv, Gunhild. 2014. Virtuous Imperialism or a Shared Global Objective? The Relevance of Human Security in the Global North. In *Environmental and Human Security in the Arctic*, ed. Gunhild Hoogensen Gjørv, Dawn R. Bazely, Maria Goloviznina, and Andrew J. Tanentzap, 58–80. New York: Routledge.

Hoogensen Gjørv, Gunhild, Dawn R. Bazely, Maria Goloviznina Bazely, and Andrew J. Tanentzap (eds.). 2014. *Environmental and Human Security in the Arctic*. New York: Routledge.

Huskey, Lee, Ilmo Mäenpää, and Alexander Pelyasov. 2014. Economic Systems. In *Arctic Human Development Report: Regional Processes and Global Linkages*, ed. Joan Nymand Larsen and Gail Fondahl, 151–183. Akureyri: Steffanson Arctic Institute.

IPCC. 2013. Summary for Policymakers. In *Climate Change 2013: The Physical Science Basis. Contributions of Working Group I to the Fifth Assessment Report of the Intergovernmental Panel on Climate Change*. Cambridge: Cambridge University Press.

IPCC. 2018. 2018: Summary for Policymakers. In *Global Warming of 1.5°C. An IPCC Special Report on the Impacts of Global Warming of 1.5°C Above Pre-industrial Levels and Related Global Greenhouse Gas Emission Pathways, in the Context of Strengthening the Global Response to the Threat of Climate Change, Sustainable Development, and Efforts to Eradicate Poverty*. Geneva, Switzerland: World Meteorological Organization.

Käpylä, Juha, and Harri Mikkola. 2015. On Arctic Exceptionalism: Critical Reflections in the Light of the Arctic Sunrise Case and the Crisis in Ukraine. FIIA Working Paper 85, The Finnish Institute of International Affairs, Helsinki.

Larsen, Joan Nymand, Peter Schweitzer, and Andrey Petrov (eds.). 2014a. *Arctic Social Indicators: ASI II: Implementation*. Copenhagen: Nordic Council of Ministers.

Larsen, J.N., O.A. Anisimov, A. Constable, A.B. Hollowed, N. Maynard, P. Prestrud, T.D. Prowse, and J.M.R. Stone. 2014b. Polar Regions. In *Climate Change 2014: Impacts, Adaptation, and Vulnerability. Part B: Regional Aspects*. Contribution of Working Group II to the Fifth Assessment Report of the Intergovernmental Panel on Climate Change, ed. V.R. Barros, C.B. Field,

D.J. Dokken, M.D. Mastrandrea, K.J. Mach, T.E. Bilir, M. Chatterjee, K.L. Ebi, Y.O. Estrada, R.C. Genova, B. Girma, E.S. Kissel, A.N. Levy, S. MacCracken, P.R. Mastrandrea, and L.L. White. Cambridge: Cambridge University Press.

Lawson, Robert J., Maciek Hawrylak, and Sarah Houghton. 2008. Human Security for an Urban Century: Local Challenges, Global Perspectives. In *Environmental Change and Human Security: Recognizing and Acting on Hazard Impacts*, ed. P.H. Liotta, David Mouat, William G. Kepner, and Judith M. Lancaster, 169–202. Dordrecht: Springer.

Lemanski, Charlotte. 2012. Everyday Human (In)Security: Rescaling for the Southern City. *Security Dialogue* 43 (1): 61–78.

Lenton, Timothy M. 2011. Early Warning of Climate Tipping Points. *Nature Climate Change* 1 (4): 201–209.

Lenton, Timothy M., Hermann Held, Jim W. Hall, Wolfgang Lucht, Stefan Rahmstorf, and Hans Joachim Schellnhuber. 2008. Tipping Elements in the Earth's Climate System. *Proceedings of the National Academy of Sciences* 105 (6): 1786–1793.

Little, Laura, and Chris Cocklin. 2010. The Vulnerability of Urban Slum Dwellers to Global Environmental Change. In *Global Environmental Change and Human Security*, ed. Richard A. Matthew, Jon Barnett, Bryan McDonald, and Karen L. O'Brien, 77–96. Cambridge: The MIT Press.

McGlade, Christopher, and Paul Ekins. 2015. The Geographical Distribution of Fossil Fuels Unused When Limiting Global Warming to 2°C. *Nature* 517 (7533): 187–190.

Meakin, Stephanie, and Tiina Kurvits. 2009. *Assessing the Impacts of Climate Change on Food Security in the Canadian Arctic.* Ottawa: GRID-Arendal.

Newman, Peter, Timothy Beatley, and Heather Boyer. 2009. *Resilient Cities: Responding to Peak Oil and Climate Change.* Washington, DC: Island Press.

Noyes, P.D., M.K. McElwee, H.D. Miller, B.W. Clark, L.A. Van Tiem, K.C. Walcott, K.N. Erwin, and E.D. Levin. 2009. The Toxicology of Climate Change: Environmental Contaminants in a Warming World. *Environment International* 35 (6): 971–986.

Nyseth, Torill. 2017. Arctic Urbanization: Modernity Without Cities. In *Arctic Environmental Modernities: From the Age of Polar Exploration to the Era of the Anthropocene*, ed. Lill-Ann Körber, Scott MacKenzie, and Anna Westerståhl Stenport, 59–70. Palgrave Studies in World Environmental History.

Obama, Barack. 2015. Remarks by the President at the GLACIER Conference. Anchorage, AK, August 31. https://www.whitehouse.gov/the-press-office/2015/09/01/remarks-president-glacier-conference-anchorage-ak. Accessed 10 Feb 2019.

Østby, Gudrun, Henrik Urdal, and Halvard Buhaug. 2013. Sustainable Urbanization and Human Security. In *A Changing Environment for Human Security: Transformative Approaches to Research, Policy and Action*, ed. Linda Sygna, Karen O'Brien, and Johanna Wolf, 56–66. London: Routledge.

Rautio, Arja, Birger Poppel, and Kue Young. 2014. Human Health and Well-Being. In *Arctic Human Development Report: Regional Processes and Global Linkages*, ed. Joan Nymand Larsen and Gail Fondahl, 299–348. Akureyri: Steffanson Arctic Institute.

Romero-Lankao, Patricia, Mercy Cordova Borbor, Eduardo Behrentz, and Laura Dawidowski. 2013. Urban Risk and Vulnerability: Insights and Lessons from Latin American Cities. In *A Changing Environment for Human Security: Transformative Approaches to Research, Policy and Action*, ed. Linda Sygna, Karen O'Brien, and Johanna Wolf, 67–82. London: Routledge.

Scherer, Glenn. 2012. Climate Science Predictions Prove Too Conservative. *Scientific American*, December 6. https://www.scientificamerican.com/article/climate-science-predictions-prove-too-conservative/. Accessed 27 Mar 2018.

Schreiber, Melody. 2018. The Race to Save Arctic Cities as Permafrost Melts. Wired.com, May 14. Accessed from https://www.wired.com/story/the-race-to-save-arctic-cities-as-permafrost-melts/.

Séguin, Jacinthe (ed.). 2008. *Human Health in a Changing Climate: A Canadian Assessment of Vulnerabilities and Adaptive Capacity*, 324–327. Ottawa: Health Canada.

Simon, David, and Hayley Leck. 2013. Cities, Human Security and Global Environmental Change. In *A Changing Environment for Human Security: Transformative Approaches to Research, Policy and Action*, ed. Linda Sygna, Karen O'Brien, and Johanna Wolf, 170–180. London: Routledge.

Smith, Laurence C. 2011. *The World in 2050: Four Forces Shaping Civilization's Northern Future*. London: Penguin.

Smylie, Janet, and Deborah Michelle Firestone. 2017. *Our Health Counts: Urban Indigenous Health Database Project*. Toronto: Tungasuvvingat Inuit and Well Living House Action Research Centre for Indigenous Infant, Child and Early Family Health and Wellbeing at the Centre for Urban Health Solutions, St. Michael's Hospital. http://www.wellivinghouse.com/wp-content/uploads/2018/04/Our-Health-Counts-Urban-Indigenous-Health-Database-Project-Inuit-Adults-July-2017.pdf. Accessed 8 Jan 2019.

Sohns, Antonia. 2017. Mining and Lack of Governance Threaten Arctic Freshwater Supplies. Fair Observer, June 7. Accessed from https://www.fairobserver.com/more/environment/mining-arctic-drilling-environmental-climate-change-global-warming-latest-news-today-54780/.

Stone, Brian. 2012. *The City and the Coming Climate: Climate Change in the Places We Live*. Cambridge: Cambridge University Press.

Studin, Irvin. 2017. Canada: Population 100 Million—Part II. *Global Brief Magazine*, March 3. Accessed from http://globalbrief.ca/blog/2017/03/03/canada-%E2%80%93-population-100-million-%E2%80%93-part-ii/.

UNDP. 1994. *Human Development Report 1994: New Dimensions of Human Security*. New York: United Nations Development Program.

US Geologic Survey. 2008. The Circum-Arctic Resource Appraisal. Accessed from https://pubs.usgs.gov/fs/2008/3049/fs2008-3049.pdf.

Wang, M., and J.E. Overland. 2009. A Sea Ice Free Summer Arctic Within 30 Years? *Geophysical Research Letters* 36 (7): 1–5.

World Bank. 2013. *Turn Down the Heat: Climate Extremes, Regional Impacts, and the Case for Resilience*. A Report for the World Bank by the Potsdam Institute for Climate Impact Research and Climate Analytics. Washington, DC: World Bank.

CHAPTER 6

Between Militarization and Disarmament: Challenges for Arctic Security in the Twenty-First Century

Heather Exner-Pirot

Abstract The period of regionalization that the Arctic has enjoyed since Gorbachev's famous Murmansk Speech in 1987 has brought stability and peace. Ironically, however, it has not led to disarmament, despite the initial premise of developing the Arctic region into a 'Zone of Peace', with an Arctic Council focused on demilitarization and arms control issues. Why have issues of militarization fallen so far off the Arctic states' political agenda? This chapter will provide an historical overview of disarmament and demilitarization discussions in the Arctic, including coordination around cleaning up nuclear and radioactive waste. It will then turn to more contemporary issues around militarization: the threat of an arms race in the region, an absence of fora to discuss issues of military security; and the geopolitical dynamics, especially concerning Russia, that make Arctic demilitarization efforts so challenging. The chapter will conclude with an assessment of the short and long-term security trends in the region and strategies for more explicit peace-building, including demilitarization.

H. Exner-Pirot (✉)
Observatoire de la politique et la sécurité de l'Arctique at CIRRICQ, Quebec, Canada

L. Heininen and H. Exner-Pirot (eds.), *Climate Change and Arctic Security*, https://doi.org/10.1007/978-3-030-20230-9_6

Keywords Arctic · Disarmament · Militarization · Confidence-building

Introduction

The period of regionalization that the Arctic has enjoyed since Gorbachev's famous Murmansk Speech in 1987 has brought stability and peace. Ironically, however, it has not led to disarmament, despite the initial but unrealized premise of developing the Arctic region into a 'Zone of Peace', with an Arctic Council focused on demilitarization and arms control issues. Why have issues of disarmament fallen so far off the Arctic states' political agenda?

This chapter will provide an historical overview of disarmament and demilitarization discussions in the Arctic. It will then turn to more contemporary issues around militarization: the threat of an arms race in the region, the current absence of fora to discuss issues of military security; and the geopolitical dynamics, especially between Russia and the West, that make Arctic demilitarization efforts so challenging. The chapter will conclude with an assessment of the short and medium-term strategies for more explicit peace-building in the Arctic (Table 6.1).

Table 6.1 Definitions of concepts

Disarmament	To reduce or withdraw weapons and military forces
Demilitarization	To remove military forces from an area; to transition to non-military means to provide for security
Denuclearisation	To remove or ban nuclear weapons from an area
Neutralization	To remove an area from existing and future military conflict
Arms Control	To mutually agree, usually with and adversary, on efforts to limit, reduce and/or control the development, production, stockpiling, proliferation and usage of weapons
Confidence-building	To conduct activities which prevent conflict, reduce tension and build trust with an antagonist
Remilitarization	To reintroduce or increase military forces to an area that had previously experienced a reduction

The Cold War and Militarization the Arctic Region

The Arctic region witnessed its first significant international conflict during World War II. Three events are of particular importance: first, the Winter War between Finland and Russia between November 1939 and February 1940, which saw Finland ultimately concede 11% of its territory to the larger and more powerful Soviet Union, despite the latter's suffering of higher casualties (300,000 compared to 65,000 for the Finns) (Andrews 2016); second, the Arctic convoys from the United States, Iceland and the United Kingdom to deliver supplies to ally Soviet Union via Arkhangelsk and Murmansk; and third the Aleutian Islands Campaign, waged between June 1942 and August 1943, in which Japan temporarily seized the small American islands of Attu and Kiska. Numerous military mega-projects, such as the Alaska Highway and the Goose Bay airport in Labrador, Canada, were established during this time.

However it was during the Cold War that the Arctic was truly established as a military theatre. This was precipitated by the spectre of nuclear war between the era's two super powers, the United States and the Soviet Union, in which the polar route provided the shortest path to major targets.

The Distant Early Warning, or DEW, Line, was one of the more significant Western responses to the Soviet nuclear threat. Social and economic impacts, disproportionately affecting Indigenous peoples, were felt across the North American Arctic from northern Alaska, across the Canadian Arctic, and into Greenland. The DEW Line was constructed between 1954 and 1957 and operational until 1985, at which time it was merged into the more modern North Warning System (Lajeunesse 2007). Cooperation on radar networks naturally extended to discussions on the integration of air defense plans, and the NORAD, or North American Air Defense Agreement, was signed in 1957 between Canada and the United States.

American air bases were furthermore established during WWII in Thule, Greenland and Keflavik, Iceland, later becoming strategic bases during the Cold War as part of the North Atlantic Treaty Organization (NATO) alliance, which both Denmark and Iceland are members of.

For its part, beginning in the 1950s the Soviet Union concentrated its nuclear arsenal in its Arctic zone, specifically the Kola Peninsula, to extend its range and put them as close to American targets as possible. The adjacent Arctic waters, including the Barents Sea, furthermore provided essential marine access to the Atlantic Ocean for the Soviet Union's Northern Fleet. It was in this way that the Arctic became a critical military zone during the Cold War, largely linked to its strategic location for nuclear strikes and deterrence.

Early Efforts to Demilitarize

There have been intermittent entreaties and proposals from academics and NGOs to demilitarize the Arctic in one way or another.

In 1964 two scientists, one American and one Soviet (Alexander Rich and Aleksandr Vinogradov), jointly proposed an Arctic nuclear free zone in the *Bulletin of Atomic Scientists*. It was revived in 1980 by Canadian academic Hanna Newcombe in the journal *Peace Research* (Newcombe 1980) and later by the Pugwash Group in August 2007 (Pugwash Group 2007) and the InterAction Council in 2011 (Axworthy 2011). There were similar proposals for nuclear free zones in the Nordic region, most recently by the Icelandic parliament in 2012 (Buckley 2013), as a response to the American withdrawal from Keflavik.

The Nordic states made recurrent efforts to disarm northern Europe across the twentieth century (see Chillaud 2006 for a comprehensive review). But the first serious effort at denuclearization was from Finnish President Urho Kekkonen in May 1963, based on the desire to keep the region, in this case Denmark, Sweden, Norway and Finland, free from speculation as to its nuclear status, given that the region was already de facto nuclear free (Broms 1989, p. 348). The Soviet Union agreed to guarantee it would not use nuclear weapons against those states as a *quid pro quo*, but Denmark and Norway, members of NATO, were dissatisfied with this arrangement and asked instead that the Soviet Union agree to include the Kola Peninsula in the NWFZ. Predictably, they declined (ibid.).

Kekkonen attempted a regional NWFZ a second time in 1978, with greater detail, but it was rejected on similar grounds. Swedish Foreign Minister, and later head of the International Atomic Energy Agency, Hans Blix, made an amendment to Kekkonen's proposal later that year, suggesting the Baltic Sea should be included in the zone, which was also rejected by the Soviet Union. The Government of Norway later suggested that a NWFZ should include parts of Central Europe and parts of the Kola Peninsula. To everyone's surprise, Soviet President Leonid Brezhnev asserted the Soviet Union was prepared to consider the inclusion of parts of the Kola Peninsula. While the discussions on the proposed NWFZ did not fail per se, they never went further (ibid.).

The Post-Cold War Arctic and Demilitarization

With the collapse of the Soviet Union in 1991 and the subsequent end of the Cold War as the United States ascended to global super power, the need for vigilance in the Arctic region lost some of its importance. Military spending dropped on all sides in the 1990s as the Cold War arms race left the Soviet Union insolvent, and non-conventional threats from e.g. terrorism and humanitarian crises assumed greater priority for the United States and NATO.

However the collapse of the Soviet military itself was a huge threat, with fears of 'loose nukes' and other weapons falling into terrorist or rogue states' hands, the threat of diffusion of weapons-related scientific expertise, and environmental impacts from improper storage, maintenance and transportation of nuclear and other weapons (see Heininen in this volume).

The now famous Murmansk speech by Mikhail Gorbachev in 1987, calling for the Arctic to become a "Zone of Peace", set the stage for how the Arctic region would look and act like post-Cold War. Gorbachev called for a "radical lowering of the level of military confrontation in the region" including the establishment of a nuclear-free zone in northern Europe; and suggested potential limitations on Soviet activities in the region, such as removing submarines equipped with ballistic missiles from the Soviet Baltic Fleet; restricting naval activity in the Baltic, Northern, Norwegian and Greenland seas; extending confidence-building measures such as observing military exercises; and eliminating nuclear testing in Novaya Zemlya (Gorbachev 1987).

The opportunity provided by Gorbachev was leveraged by the Arctic's smaller powers, and by Finland in particular with the establishment of the Arctic Environment Protection Strategy (AEPS) in 1991. As Mullen (1995) describes it, during AEPS negotiations the Nordic countries were interested in addressing the threat of radioactive contamination and controlling nuclear activities in the Arctic (pp. 29–30). However the United States resisted the movement of disarmament talks to regional-level fora and the USSR was "very sensitive" to references in the AEPS Declaration to either nuclear contamination or nuclear weapons testing (p. 48). It was later revealed that Russia was dumping nuclear reactors and radioactive waste off the coast of Novaya Zemlya, and they eventually conceded that they needed assistance to deal with the waste in other ways (p. 49). An Arctic Military Environmental Cooperation (AMEC) project was eventually established in 1996 by Russia, Norway and the United States, and joined in 2003 by the United Kingdom, to address military-related environmental challenges in the Arctic.

While the topic of Arctic nuclear environmental waste was being carefully managed in AEPS negotiations, Canada was beginning to lobby for the creation of a more formal intergovernmental forum with a broader mandate: what would eventually be the Arctic Council. The domestic roots of Canadian interest in Arctic regional cooperation largely stem from disarmament activists. In June 1987, four months before Gorbachev's Murmansk speech and one week before Ronald Regan's famous entreaty to Gorbachev to "tear down this wall" (referring to the Berlin wall), Canada released a Defence Review that called for the acquisition of 10–12 nuclear attack submarines to patrol the waters of the Arctic (it was later cancelled). This sparked immediate opposition from northern Indigenous leaders, including Mary Simon, the president of the Inuit Circumpolar Council at the time, as well as environmental groups. Neither Moscow nor indeed Washington DC were enthused with the proposal. Gorbachev's speech a few months later provided an impetus for certain academic, Indigenous, environmental and philanthropic groups to formally advocate for a 'zone of peace' through regional disarmament in the Arctic (see English 2013), first within Canada and then more broadly amongst the Arctic states. The lobbying proved successful enough to encourage Canadian Prime Minister Brian Mulroney to propose a Council of Arctic countries in Leningrad in November 1989, though he avoided mentioning security or disarmament specifically, due to American sensitivities on the subject. Simon was later made Canadian Circumpolar Ambassador and spearheaded the discussions from the Canadian side.

American unwillingness to discuss security, regional disarmament or demilitarization nearly derailed the Arctic Council idea completely. Despite its origins in Gorbachev's 'zone of peace' concept, the Canadian initiative to establish an Arctic Council ultimately evolved from a regional forum to address security and disarmament, to beefing up the more technically-oriented AEPS to become a political forum that addressed sustainable development as well as environmental protection issues. The major foci of negotiations for the Arctic Council became instead the extent to which Indigenous organizations should be formally involved (ibid.). The 1996 Ottawa Declaration, on The Establishment of an Arctic Council, encapsulated this evolution with a footnote, iterating that "the Arctic Council should not deal with matters related to military security". However it should not be forgotten that the original genesis of regional-level collaboration in the Arctic was around disarmament.

A 'New' Cold War: Remilitarization of the Arctic?

The 1990s and early 2000s were marked by the United States' unipolar moment, concomitant with a reduction in conventional forces. This allowed for a focus on human security issues, including environmental, economic, and cultural ones, not least in the Arctic which, due to a paucity of either voters or economic weight, lent itself to an idealism generally found lacking in more populated areas.

Another decade went by with circumpolar affairs falling relatively low on the priorities of defence and foreign ministries. But a series of events had changed the political calculation on the Arctic by 2008.

The first was the launching of the 2004 *Arctic Climate Impact Assessment* by the Arctic Council and the International Arctic Science Committee, which proved definitively that climate change was not only possible, but was already occurring with dramatic effect in the Arctic region. The fact that the eight Arctic states, including the United States under George W. Bush, acknowledged and endorsed the ACIA and its findings in the 2004 Reykjavik Declaration, lent it further credibility. States and corporations were now fully aware that warming in the Arctic was melting sea ice, which could make both extraction and transportation of resources more economically feasible.

The second was the infamous planting of the Russian flag at the seabed of the North Pole by a Russian-sanctioned but privately led expedition in August 2007, which raised concerns about a "scramble" or "race" for the extended continental shelf in the Arctic. The previous month, Canadian Prime Minister Stephen Harper had held a media event in the Arctic, declaring that "Canada has a choice when it comes to defending our sovereignty in the Arctic, we either use it or lose it" and subsequently announced billions of dollars in military spending for the region. The combination provoked media headlines and magazine covers.

Finally, in 2008 the United States Geological Survey made a high profile and ubiquitously cited appraisal of the Arctic region's hydrocarbon reserves, estimating it at 90 billion barrels of oil, 1669 trillion cubic feet of natural gas, and 44 billion barrels of natural gas liquids (USGS 2008, p. 4), at a time when oil prices were at record highs.

The combination of access, availability and competition prompted all eight Arctic states to develop Arctic strategy documents, detailing their interests, activities and ambitions for the region, in the subsequent five years (Heininen 2012), including (re)investments in Arctic defense capabilities (for an inventory of Arctic military assets see Regehr and Jackett 2018).

Despite the recent commodities bust, strategic interest in the region has not subsided. In 2018 both NATO and Russia ran their largest ever Arctic exercises: Trident Juncture and Vostok-2018, respectively; and there are obvious tensions between Russia and the West over Russia's more assertive and revisionist foreign policy in the past decade including incursions into Georgia in 2008 and Crimea in 2014, and electoral interferences in the United States and elsewhere. Together, these factors have perpetuated concerns of a new Arctic Cold War or arms race.

Analysts differ as to the motivations, degree and impact of the re-militarization of the Arctic. While hawks on both sides are concerned about the other side's intentions, all Arctic states have publicly asserted and conceded that there is no risk or rationale for a conventional military conflict in the Arctic. Recent military investments have common and logical underpinnings. The first is political; for several Arctic states, the region is tied deeply to their national identity and for Canada, Norway and Russia in particular, Arctic defense is an opportunity to project pride and strength—a winning domestic political strategy.

The second is practical. The Arctic is warming and the sea ice is melting. This has precipitated increased shipping and other activity. While conditions are still unpredictable and shipping traffic is still very low compared to other global shipping routes, there is a need for states to be able to exert greater stewardship and control over the Arctic, including for constabulary duties, scientific research, environmental monitoring and enforcement, and search and rescue.

The third is strategic. Defence analysts in both the West and Russia profess a need to project force in the region, if only for deterrence reasons. As Knut Hauge, the director general of the Security Policy Department at the Norwegian Ministry of Foreign Affairs has articulated, "it is very, very important to show the necessary firmness, or you can call it deterrence, towards the Russians…You can say that a prerequisite for being able to engage with Russians as a small country in our neighbourhood is that we have the backing of our allies and that our defence is credible" (as quoted in Sevunts 2018).

With resources, territory, and Cold War cleavages at play, the assertion of realist principles, including the need for military resources, has found a ready audience in Arctic governments.

All of this leads to a fundamental discussion: can one objectively state that the Arctic is remilitarizing? If so, to what levels? Has it met or exceeded Cold War levels, and if so by what measure? While it is clear military investment has risen from the levels of the late 1990s and early 2000s, there is no available quantitative analysis indicating whether it is disproportionate to overall military spending amongst the Arctic states. This remains a gap not least because defining what is Arctic-specific military spending and what is general security spending would ultimately be a subjective exercise.

Confidence-Building Efforts in the Twenty-First Century

Although disarmament per se has fallen off political agendas, in the Arctic and elsewhere, there have been constructive confidence-building efforts in the region since 2008. A first, political, effort at Arctic confidence-building in the post, post-Cold War period was the convening of the so-called 'Arctic 5'—the littoral Arctic states including Canada, Russia, Norway, Denmark the United States—in Ilulissat, Greenland in May 2008. High level representatives affirmed their commitment to the "extensive international legal framework" that applies to the Arctic Ocean, and to "strengthen this cooperation, which is based on mutual trust and transparency" in the region (Ilulissat Declaration 2008).

The states further committed in Ilulissat to "strengthen search and rescue capabilities and capacity". Three years later, the eight Arctic states concluded an *Agreement on Cooperation on Aeronautical and Maritime Search and Rescue in the Arctic* under the auspices of the Arctic Council, during the 2011 Nuuk Ministerial. Although the Agreement itself contained few new obligations on states and was widely based on previous international SAR conventions (Exner-Pirot 2012), it provided space and political impetus to better coordinate what was already possible legally, e.g. around sharing information, arranging visits between key personnel and sharing personnel and equipment during rescue operations

(Agreement on Cooperation on Aeronautical and Maritime Search and Rescue in the Arctic 2011, p. 6). The defense apparatuses of the Arctic states took advantage of this opening and established an Arctic Coast Guard Forum in November 2015. The forum, including all eight Arctic states, meets twice annually. Although it is an independent body and not bound or bolstered by treaty, its chairmanship mirrors the Arctic Council's. It held its first live exercise in Reykjavik in September 2017 and its second in Finland in April 2019.

The Arctic Coast Guard Forum is the most successful current example of confidence-building in the Arctic, for the simple reason that it is still active. There have been several other recent initiatives that have been suspended or diminished due to the Crimean conflict. Russia and Norway held the POMOR joint exercises from 2010 until 2013, after which cooperation was suspended. Similarly, FRUKUS naval exercises between France, the United Kingdom, the United States and Russia were held from 2003 until 2013. An Arctic Security Forces Roundtable was initiated with Russia, the other seven Arctic nations, and several NATO allies in 2011 but has excluded Russia since 2014. And the eight nation Arctic Chiefs of Defense Staff met only twice, in April 2012 and June 2013, until it was suspended.

Some relics of cooperation still exist; the Norwegian Joint Headquarters in Bodø still has a 24/7 operative hotline to its colleagues at the Northern fleet in Severomorsk (Gronning 2017). Furthermore Russia sent observers to NATO's Trident Juncture in October/November 2018, and NATO member Turkey sent observers to Vostok-2018. But confidence-building efforts have certainly been diminished.

A Way Forward: Confidence-Building, Compartmentalization and Neutralization

The current militarization of the Arctic, to the extent that it exists, is concentrated in conventional forces. All states assert that they are using these resources either defensively or in a constabulary capacity given the increase in economic and other activity in the region. There have been consistent and clear messages from both Western and Russian leaders that interests in the Arctic are shared and that it is in everyone's best interest to maintain regional collaboration, in particular within the Arctic Council. A selection of quotes from the latest Arctic Council Ministerial in Fairbanks, in May 2017, highlight the common sentiment:

> The Arctic Council "an indispensable forum in which we can pursue cooperation." (U.S. Secretary of State Rex Tillerson)
>
> In the Arctic "there is no potential for conflict." (Russian Foreign Minister Sergei Lavrov)
>
> The Arctic Council "is a place where we the Arctic nations can set aside issues outside the Arctic, and appreciate that we have the real honour of shared stewardship of this incredibly beautiful and precious region." (Canadian Foreign Minister Chrystia Freeland)
>
> "Together we are not only making a major contribution to stability and prosperity in the Arctic. We are turning the Arctic into a global model for peace and sustainable development." (Norwegian Foreign Minister Børge Brende)

Given the commitment amongst Arctic states to promote and maintain peace and stability in the Arctic region—a genuine desire shared by all eight—what efforts could be pursued to reduce the potential for military conflict?

Denuclearisation Unlikely

There is precedence for nuclear weapons free zones, with seven declared around the world, including the southern pole; Antarctica. Norway's Spitsbergen archipelago is furthermore demilitarized as a result of the 1920 Spitsbergen Treaty (see Buckley 2013).

However total denuclearisation in the Arctic would be very difficult politically. As articulated above, Russia has a stockpile of nuclear weapons located in the Kola Peninsula. The region and its assets are core to Russia's nuclear deterrence strategy. Because it is weaker than the United States and the NATO alliance, in terms of conventional military force, its nuclear assets are critical to its security, as well as its status as a great power. Denuclearisation would also be counter to President Vladimir Putin's political brand of leading a resurgent, strong Russia.

Neither would the United States, the only other nuclear power in the Arctic region, be interested in constraining its nuclear forces range, with its air base in Fairbanks, radar station in Thule, and ballistic missile submarines.

This is aggravated by the current political situation, with US President Donald Trump announcing on October 20, 2018 his intention to terminate the 1987 Intermediate Range Nuclear Forces (INF) Treaty (eliminating nuclear and conventional ground-launched ballistic and cruise missiles with ranges of 500–5500 kms) due to allegations that Russia has been in violation of the treaty since 2014 (Arms Control Association 2018). It is unlikely in this context that new nuclear disarmament talks would begin anytime soon.

Confidence-Rebuilding

By contrast, the ground is already prepared in the short term for enhanced Confidence and Security-Building Measures (CSBMs). OSCE characterizes these as the exchange of military information; verification of compliance with agreed commitments; and different forms of military cooperation which "aim to reduce the risk of conflicts, increase trust… and contribute to greater openness and transparency in the field of military planning and military activities" (as quoted in Schaller 2014, p. 7). The Arctic Coast Guard Forum is an excellent model. While its focus is constabulary rather than defensive, it still provides an excellent trust building opportunity for the Arctic states.

However it is time to reconvene the Arctic Chiefs of Staff meetings as a constructive and important form of defense diplomacy. Certainly Russia deserves consequences for its aggression against other states, and joint exercises may require some more time. But meeting again at the rank of Chiefs of Staff would enhance the Arctic states' mutual trust and ability to communicate at multiple levels in the event of an incident that escalated regional tensions.

Such efforts would involve "compartmentalizing" Arctic strategies from broader defense policy ones. Compartmentalization refers to the recent practice of Arctic states to evaluate and pursue cooperation on Arctic-specific issues, especially between Russia and the West, separately from their other dealings. The Arctic Coast Guard Forum is an example of this—the willingness to go ahead with paramilitary collaboration even in the wake of Russia's Crimean incursion and the decision to suspend almost all other military cooperation. Some scholars have referred to this as an Arctic exceptionalism—where the region is treated as a regional security complex with its own, independent, political calculus that is poorly explained by conventional realist theories of international relations (see Exner-Pirot and Murray 2017; Heininen 2016). The fact is the Arctic states have primarily converging interests in the region and all benefit from cooperation and stability. Maintaining this cooperation requires compartmentalizing the region from broader geopolitical considerations at times.

Neutralization

In the medium term, militarization, disarmament and denuclearization are unlikely to be achieved, let alone initiated by the Arctic states. But there is another option which would reduce the prospect of actual conflict in the region: neutralization, or the removal of the Arctic from the field of possible war.

Neutralization is a concept that arose in popularity in the nineteenth century, with Switzerland for example having been declared a neutral state since the Congress of Vienna in 1815. The Panama Canal has also been declared neutral. Austria, Finland and Sweden are formally neutral states.

More relevant to the Arctic region, the International Space Station and Antarctica are areas where the United States, Russia and others have interests but have been excluded as theatres of potential conflict.

For the Arctic, a formal declaration of neutralization of the Central Arctic Ocean on behalf of the Arctic Five—an Ilulissat 2.0—would simply makes explicit what is already the practice. All of the Arctic states, in their respective strategies and in public declaration by their leaders, have asserted that they neither expect, nor would they initiate, military conflict over Arctic territory.

Neutralization does not require the removal of defensive and constabulary capabilities; only a commitment to maintain the peace and stability that exists in the region. It is not a perfect solution—a region is only neutral until it is not—but it would go a long way towards formalizing what the Arctic states have already expressed, and help change the dominant media narrative and public perception of an Arctic that is experiencing military build-up.

Conclusion

One of the defining features of the Arctic region is its peacefulness and stability, despite rapid environmental and political changes since the 1980s (Heininen 2011, p. 40). Many political leaders, activists, and academics have attempted to formalize the Arctic as a zone of peace through denuclearisation and demilitarization, to no avail, and advocacy and debate on the subject has dwindled. As climate change imposes a paradigm shift for the region, is there any opportunity to rekindle disarmament efforts?

This chapter argues that the time is ripe for enhanced confidence building measures between Russia and the other Arctic states. It is time to reinstate the many positive initiatives that were suspended after the Russian incursion into Crimea in 2014, including the Arctic Chiefs of Defense Staff meeting and joint exercises.

It would furthermore benefit all Arctic states, and their stated goals of peace and cooperation in the region, to formally declare the Arctic as a neutralized area—off limits from war. This would not require active disarmament or denuclearisation, but would help enhance trust amongst all parties and shift the narrative of an Arctic on the brink of conflict.

The contemporary Arctic region has been shaped by speech acts—from Gorbachev in Murmansk in 1987 calling for a zone of peace and various disarmament measures; to the Arctic Five Ministers in Ilulissat in 2008 committing to the rule of law, the strengthening of cooperation and mutual trust and transparency; to the Arctic Council Ministerial declarations. It is time for a further declaration of the Arctic states' intentions, of the Arctic as a neutralized area. Regional peace is not a given; it requires nurturing.

References

Agreement on Cooperation on Aeronautical and Maritime Search and Rescue in the Arctic. 2011. Arctic Council, May. Accessed from https://oaarchive.arctic-council.org/bitstream/handle/11374/531/EDOCS-3661-v1-ACMMDK07_Nuuk_2011_SAR_Search_and_Rescue_Agreement_signed_EN_FR_RU.PDF?sequence=5&isAllowed=y.

Andrews, Evan. 2016. What Was the Winter War? *History Channel*, November 30. Accessed from https://www.history.com/news/what-was-the-winter-war.

Arms Control Association. 2018. The Intermediate-Range Nuclear Forces (INF) Treaty at a Glance, December. Accessed from https://www.armscontrol.org/factsheets/INFtreaty.

Axworthy, Thomas. 2011. A Proposal for an Arctic Nuclear Weapon-Free Zone. InterAction Council. Accessed from https://www.interactioncouncil.org/publications/proposal-arctic-nuclear-weapon-free-zone.

Broms, Bengt. 1989. Proposals to Establish a Nordic Nuclear-Weapon-Free Zone. *Michigan Journal of International Law* 10 (2): 345–361.

Buckley, Adele. 2013. An Arctic Nuclear-Weapon-Free Zone: Circumpolar Non-Nuclear Weapons States Must Originate Negotiations. *Michigan State International Law Review* 22 (1): 167–194.

Chillaud, Matthieu. 2006. Territorial Disarmament in Northern Europe: The Epilogue of a Success Story? SIPRI Policy Paper 13, August. Accessed from https://www.sipri.org/sites/default/files/files/PP/SIPRIPP13.pdf.

English, John. 2013. *Ice and Water: Politics Peoples and the Artic Council.* Toronto: Penguin Canada.

Exner-Pirot, Heather. 2012. Defence Diplomacy in the Arctic: The Search and Rescue Agreement as a Confidence Builder. *Canadian Foreign Policy Journal* 18 (2): 195–207.

Exner-Pirot, Heather, and Robert Murray. 2017. Regional Order in the Arctic: Negotiated Exceptionalism. *Politik* 20 (3): 47–64.

Gorbachev, Mikhail. 1987. Speech in Murmansk at the Ceremonial Meeting on the Occasion of the Presentation of the Order of Lenin and the Gold Star to the City of Murmansk, October 1. Accessed from https://www.barentsinfo.fi/docs/Gorbachev_speech.pdf.

Gronning, Ragnhild. 2017. The Need for Confidence-Building Measures in the North Is Greater Than Ever. *High North News*, December 7. Accessed from https://www.highnorthnews.com/en/need-confidence-building-measures-north-greater-ever.

Heininen, Lassi. 2011. The End of the Post-Cold War in the Arctic. *Nordia Geographical Publications* 40 (4): 31–42.

Heininen, Lassi. 2012. State of the Arctic Strategies and Policies—A Summary. In *Arctic Yearbook 2012*, ed. Lassi Heininen, Heather Exner-Pirot, and Joel Plouffe. Akureyri, Iceland: Northern Research Forum.

Heininen, Lassi. 2016. *Future Security of the Global Arctic: State Policy, Economic Security, and Climate.* New York: Palgrave Macmillan.

Ilulissat Declaration. 2008. Declaration from the Arctic Ocean Conference, May 28. Accessed from https://cil.nus.edu.sg/wp-content/uploads/2017/07/2008-Ilulissat-Declaration.pdf.

Lajeunesse, Adam. 2007. The Distant Early Warning Line and the Canadian Battle for Public Perception. *Canadian Military Journal* 8 (2): 51–59.

Mullen, Patrick. 1995. State Interests and the Arctic Environmental Protection Strategy. Dissertation. Accessed from https://search.proquest.com/docview/304200051?pq-origsite=primo.

Newcombe, Hanna. 1980. A Proposal for a Nuclear-Free Zone in the Arctic. *Peace Research* 12 (4): 175–181.

Pugwash Group. 2007. Canadian Pugwash Call for an Arctic Nuclear-Weapon Free Zone, August 24. Accessed from http://www.pugwashgroup.ca/events/documents/2007/2007.08.24-Arctic_%20NWFZ.pdf.

Regehr, Ernie, and Michelle Jackett. 2018. Circumpolar Military Facilities of the Arctic Five, July. The Simons Foundation. Accessed from http://www.thesimonsfoundation.ca/sites/default/files/Circumpolar%20Military%20Facilities%20of%20the%20Arctic%20Five%20-%20updated%20to%20July%202018_3.pdf.

Rich, Alexander, and Aleksandr P. Vinogradov. 1964. Arctic Disarmament. *Bulletin of the Atomic Scientists* 20 (9): 22–23.

Schaller, Benjamin. 2014. Confidence- and Security-Building Measures in the Arctic: The Organization for Security and Co-operation in Europe as a Role Model for the Area? In *Arctic Yearbook 2014*, ed. Lassi Heininen, Heather Exner-Pirot, and Joel Plouffe. Northern Research Forum. Accessed from https://arcticyearbook.com/images/yearbook/2014/Scholarly_Papers/23.Schaller.pdf.

Sevunts, Levon. 2018. NATO's Arctic Dilemma. *Eye on the Arctic*, December 3. Accessed from http://www.rcinet.ca/eye-on-the-arctic-special-reports/norway-nato-trident-juncture-exercise-arctic-kirkenes-russia-military-defence-tensions/.

United States Geological Survey. 2008. Circum-Arctic Resource Appraisal. Accessed from https://pubs.usgs.gov/fs/2008/3049/fs2008-3049.pdf.

CHAPTER 7

Before Climate Change, 'Nuclear Safety' Was There—A Retrospective Study and Lessons Learned of Changing Security Premises in the Arctic

Lassi Heininen

Abstract This chapter is about the environment as a special feature of Arctic geopolitics and security. The focus is on (long-range) pollution, as the first trigger of global changes in the Arctic, and 'environmental awakening' as a consequent reaction. This is also a retrospective study on 'nuclear safety' in the European Arctic, as a part of 'environmental awakening' and environmental politics in the North, and how it caused a shift in security premises of the Arctic states. Discourses of critical security studies—the risk society theory and ecological security—are applied to broaden our understanding of narrow traditional—national, competitive, military—security towards comprehensive one, as well as to analyze and illustrate changes in both Arctic security nexus and Arctic geopolitics.

Keywords Environment · Security · Security premises · Pollution · Nuclear safety · Environmental awakening

L. Heininen (✉)
INAR at University of Helsinki, Helsinki, Finland
e-mail: lassi.heininen@helsinki.fi

L. Heininen and H. Exner-Pirot (eds.), *Climate Change and Arctic Security*, https://doi.org/10.1007/978-3-030-20230-9_7

The interdependence between the environment, development and security/peace (disarmament) was originally defined in various United Nations reports (e.g. UN 1982, 1983). Since then, interrelations between climate, energy and development have been reported by scientific research, in particular the reports of the Intergovernmental Panel on Climate Change (IPCC). In today's world these interrelations, together with the societal dimension, form an important nexus: the climate is dependent on how (much) resources and energy, especially hydrocarbons, are used, since emissions from the energy sector represent roughly two-thirds of all anthropogenic greenhouse-gases. Fossil fuels, when supporting modernization (artificial) economic growth and the military, contribute to climate warming and pollute the environment as human impacts on rapid climate change. Following from this, environmental and climate policies have become parts of high-level global politics aiming to reach the goals of environmental protection and climate change mitigation, for example by developing more efficient energy technologies, promoting cooperation for low-carbon and clean energy sources, and aiming to search for a paradigm shift in security.

Climate change, first-of-all rapid global warming, became an 'issue' and problem to start to dominate discussions during the early twenty-first century, much due to several scientific reports, including the United Nations Framework Convention on Climate Change (UNFCCC) process with Conference of the Parties (COPs), the Nobel Peace Prize, and other public acts. In its reports, the IPCC has concluded that in the absence of ambitious commitments and urgent action, climate change will have severe, irreversible impacts across the world. The target of keeping the increase of long-term average temperatures below two degrees Celsius relative to pre-industrial levels, requires substantial and sustained reductions in global emissions. Even if the energy intensity of the global economy continues to decline, at current rates it would not be enough to limit the rise of global temperature.

Thus, climate change is considered to be, on the one hand, a 'wicked' problem with its direct and indirect impacts and the trigger for rapid warming of the Arctic region. Even so, and despite the Paris Agreement being in force, the "carbon tipping point for good" has passed. This leads one to ask whether the Agreement will make a difference as such: Would mitigation solve the problem, and/or are we able to adapt to big changes; or is the goal simply too little, too late? On the other hand, climate change can be interpreted to be a game-changer when it comes to security and how to define security (e.g. Kalliojärvi in this volume). Behind this is the related uncertainty meaning that its impacts may turn the environment into a less favorable place to pursue industrial activities, and cause serious consequences.

Following from this, it is understandable that climate change has become a factor strongly influencing security discourses and premises, and has been interpreted as a major trigger of changes in geopolitics and security. This is very much the situation in the Arctic in the twenty-first century, as the following examples show:

First, as a part of the bigger picture of global warming there is a/the 'Arctic Paradox', i.e. the faster we use fossil fuels, the sooner we get access to new oil and gas resources (see also Palosaari in this volume). The paradox (climate change and mass-scale exploitation) reveals that the '*Anthropocene*' is at play in the Arctic. This is a potential tipping point for the entire Earth system, and means a need for another shift in security premises. As well as, there is an emerging threat: more conflicts, and less room for resilience.

Second, due to the fact that energy in the Arctic context is usually understood as synonymous with oil and gas production mostly for foreign export "...any societal aspects associated with energy-related developments remain largely neglected" and were left into a margin in energy debates (Lempinen 2017, p. 5). This much supports the idea and discourse of the keen interrelations between the environment, climate and energy, and peoples and societies in the Arctic.

Third, the Arctic plays a key role in global ecosystem and bio-geophysical processes, which correspondingly are heavily impacted by climate change and other global changes. These have been integrated with current global economics, corresponding energy security dynamics, and finally, are related to world politics. Further, a rapid global warming, together with new multi-dimensional dynamics, has made Arctic geopolitics global, and the new Arctic geo-physical and socio-economic dynamics affects worldwide the Earth. Recent research considers that the globalized Arctic has significant implications worldwide, and has become part and parcel of global multi-dimensional change(s) (Heininen and Finger 2017).

Fourth, as wicked problems, unavoidable and rapid global warming and environmental degradation require global action. This manifests a need for the critical approach of security and geopolitics which consists—not only of resources and energy, states and (military) power—but also of the environment and climate, and knowledge and ethics as factors (of empowerment), as well as the non-state actors as subjects to decide how land, water, resources and energy are used, and how the environment is treated.

Climate change is not, however, the first trigger, since there in the Arctic was something previously alarming causing peoples' concern about the state of the environment. There was long-range—air and water—pollution impacting the region's fragile ecosystem and its peoples' and societies' everyday lives. In particular 'nuclear safety' became a trigger for a need to redefine environmental security and insert into consideration a shift in security premises in the Arctic region. Therefore, the environment was put into focus, in rhetoric, when the Arctic Environmental Protection Strategy (AEPS) was signed in 1991 (The Rovaniemi Declaration 1991). This was continued by establishing the Arctic Council in 1996, when the Arctic states affirmed their commitment to "*the well-being of the inhabitants, …sustainable development in, …the protection of the Arctic environment*" (The Ottawa Declaration 1996).

Following from this, 'environmental protection' was taken as the main area of international Arctic cooperation, and thus the environment was taken as a special feature of Arctic geopolitics. All this was before climate change was interpreted as the 'trigger' of global changes, and a new security factor, which has much influenced and materialized ecological/environmental security discourse in the Arctic.

The focus of this chapter is on pollution as the first trigger of global changes in the Arctic, as well as on the environment as a special feature of Arctic geopolitics and security. Briefly saying, that in the Arctic context—among Arctic communities and peoples—the environment matters! This is also a retrospective study and analysis on 'nuclear safety', as a part of environmental politics and the first target of 'environmental awakening' in the Arctic, in particular in the Barents Sea area, as well as how it caused a shift in security premises of the Arctic states. Discourses of critical security studies, such as the risk society theory and ecological security, are applied here on the one hand, to broaden our understanding of narrow traditional—national, competitive, military (e.g. Newcombe 1986)—security to comprehensive one; and on the other hand, to manifest an urgent need for new security agendas. Finally, there might also be lessons-to-learn, if the Arctic's security nexus should, or would, be reconceptualized.

Theoretical Background

Pollution (here trans-boundary and long-range pollution) and climate change are global issues, and closely tied with modernization and industrialization. They are the main reasons for the so-called 'environmental awakening' by citizens and the related discourse on common/comprehensive security by UN reports (e.g. UN 1983). All this resulted in the 'politicization' of the environment, which has been, and is still, cumulative, even exponential.

Environmental awakening started in the 1960s due to environmental degradation by chemicals and other toxics, as well as new kinds of environmental catastrophes, from industrial factories and nuclear accidents. It spread all over the western 'First' World, and reached the entire North in the 1980s. It was all about peoples' concern on a state of the environment due to growing needs and demands for natural resources, much due to rapid population growth and rising living standards. Pollution, in particular chemical components and radioactive materials, was the main factor causing 'environmental awakening' pushed by citizens and groups of citizens, environmental movements and other civil organizations. Protests and other activities pushed governments to act, and as an outcome a chain reaction started (Heininen 2013).

Behind the 'environmental awakening', as well as 'politicization' of the environment, was not only pollution, but also modern ecological crises, such as the Chernobyl nuclear power station accident (in 1986), and several nuclear submarine accidents (in 1980–2000), as well as the related observation of a risk society. In particular, nuclear accidents demonstrated the multiple character of ecological crises and the vulnerability of our modern, technology-driven and -dependent societies with bigger risks, as Beck (e.g. 1992) argues in his risk society theory. There is a principle difference between a threat and a risk: A threat can be physical, social or economic, and is mostly subjective and psychological. It is based on the perception of the seriousness of perceived threats in a public consciousness of a society (Haila and Heininen 1995). Unlike a threat, it is possible to qualify, rank and measure a risk-based on quantative analysis of a probability calculation. Because risks are taken as consciously, they are present in our modern "risk society", even if a risk is interpreted to be real it affects your behaviour, whether it materializes or not.

However, the 'ecology' damaged by an accident does not happened "out there but within the society", and it is impossible to define ecological problems without 'implicitly'. Therefore, "the environment cannot be isolated within a specific policy field" but it has implications for all (Haila and Heininen 1995). This includes an idea of social order, and if "social order first", then there are strict environmental regulations and control. This means theoretically that ecology has become, or has been accepted, as a "new discipline for disciplining". Of course, you should ask "what kind" of order, since "no solution to ecological problems once and for all… solidarity is needed" (ibid.).

Climate change is a global, 'super wicked' problem due to, among others, the following reasons: It goes through the entire society—locally and globally—and their fields and actions, and heavily impacts peoples' lives. Further, time is running out and risks are increasing. There is no coordinated global governance system(s) addressing the rapid climate change. The current policies seem to favor present consumption (over possible future gain). In addition to the fact that time is running out, risks are increasing, and the sea ice is melting, there are other hindrances, though not necessarily obstacles. Environmental degradation and unavoidable global warming as wicked problems, and their combination as a super wicked problem, require global action and governance. They are, however, difficult to solve because of incomplete, contradictory, and changing requirements. The related uncertainty of climate change means an emphasis on food safety and has brought both droughts and furious storms, masses of environmental refugees and asylum seekers and influential populist movements, as well as growing economic inequality between the Elites and the Masses based on NASA's HANDY Project (Nafeed 2014): 1 billion poor vs. 1% greedy, tax havens, polarization, populism, anti-elitism, riots, etc. However, no coordinated global governance system(s) addressing the rapid climate change—in-spite of oil spill preparedness/response technology/cleantech/"the best stewardship"—there is neither an Arctic Treaty nor a legally-binding treaty on climate change. Current policies seem to favor present consumption—as well as a modernization and industrialization of the Arctic.

Environmental awakening was, and is still, a moral protest against the faith of economic growth and of technology as a solution for everything. Behind is a philosophy that what is not good for people cannot be good for a nature. This wisdom is currently challenged by the 'Anthropocene'—the most recent stage of geological evolution, where industrial civilization has itself become a geological force (Finger 2016), which is at play, as heavy metals, radioactivity, and plastic in nature well show. However, it is good to remember that due to the 'awakening' the Environment, as a new term/concept, was reinterpreted to mean "the material basis for human existence, which is in a danger to be destroyed as a result of human activities" (Haila 2001). Thus, the philosophy seems actually to be more anthropocentric, as it is said to be for nature conservation.

Followed from this, environmental issues became a good reason for states to take into consideration the environment and make laws and legislation, as well as international treaties, for environmental protection. Already during, and even more after, the Cold War period several states interpreted large—international or global—problems, which cross national borders and cause risks of pollution, as part of (new) foreign policy. These included security problems, such as proliferation of nuclear weapons and terrorism, and environmental problems, such as long-range pollution and radioactive contamination. There have been several processes of international negotiations, and many international treaties, on nuclear security aiming at a reduction of nuclear risks coming from nuclear weapons and nuclear power plants: The first international agreements on trans-boundary pollution were signed in the second half of the twentieth century, such as the Oslo Dumping Convention and the London Dumping Convention of 1972 (not to dump radioactive material into a sea), which was strengthened in 1993 with a permanent ban on the dumping of radioactive waste and industrial waste.

The advanced technology has not, however, meant a decrease of risks; on the contrary, it became the ultimate reason for bigger risks and more severe catastrophes due to nuclear accidents, either caused by human mistake or technical error, or a combination of these two. No wonder that 'nuclear safety' became another main factor of 'environmental awakening' (chemical compounds being the other). As a part of the modernization and industrialization trend, as well as the advanced military technology, after the 2nd World War nuclear power promised cheap and safe energy. Thus, 'nuclear safety' much became a symbol of a new kind of modern society. Finally, it was 'nuclear safety' due to those accidents and radioactive wastes, which was behind the 'northern environmental awakening'.

It also connects the 'environmental awakening' to comprehensive security as a new security discourse, since nuclear energy was originally developed for nuclear weapons, a new mass-destruction weapon to destroy and kill. Behind this was, and much still is, a blind faith in technology, even so that being at the edge of traditional decision-making is still based on the belief that 'risk technology' will save us. There is also evidence that under the umbrella of secrecy and national security interests (e.g. the secret agendas of the major nuclear weapon states) military systems have caused, and still cause, environmental damages on a scale that would not be acceptable within other sectors of a state.

A part, although not mainstream, of the comprehensive security discourse is the relevant relationship between environmental degradation and security matters (Galtung 1982). Indeed, there are relations, as well as contradictions, between the environment and the military, which can be seen already in the routine activities in peace time of how armies use and pollute the air, water and land, utilize natural resources and cause greenhouse gasses (Heininen 2000). If pollution and other environmental issues have also a security dimension, several states have declared climate change as a national security threat, for example in the context of the UN Security Council (see Kalliojärvi in this volume). This means that rapid, uncontrolled climate change with uncertainties threatens peoples' everyday security and violates human rights (e.g. UN 1994).

Correspondingly, the discourse on common/comprehensive security, including the aspects of the environment, societies and human beings, emphasizes the point of view of peoples and their needs. One of the aims was to replace the old-style thinking of ***traditional security*** which deals with the core of unified states based on weapon-oriented security, finally guaranteed by the military. Unlike traditional security ecological/environmental security, as well as human security, directly deal with the environment, and individuals and societies and their important issues. The latter ones—health, food safety, social and economic conditions, self-determination, freedom from fear, as well as a nature—were driving forces for the 'environmental awakening' in the North.

Retrospect to Environmental Awakening and Nuclear Safety in the Barents Sea Area[1]

Though not mentioned that often any more, it was long-range air and water pollution which was interpreted as the first trigger for environmental 'awakening' in the Arctic region: DDT, POPs, heavy metals, Arctic haze, in particular radioactive materials (e.g. from Chernobyl nuclear power station) from lower latitudes of the northern hemisphere as the first grand environmental challenges to the northern-most ones

(AMAP 1997). Following from this, it became obvious that the entire North had become a primary source of pollutants and emissions. Pollution, and the threat and risks of radioactive contamination, caused protests and campaigns across national borders against nuclear weapons and wastes, which then became an international movement with peace and environment rallies and festivals (Heininen 2013). Among protesters and organizers of these protests, campaigns and rallies were environmental movements and other civil organizations, INGOs, Saami people, and scientists and scholars, who became concerned about the state of the environment due to pollution.

It was the Barents Sea area, the case study of this chapter, which was the risky hotspot of nuclear safety in the end, and in particular just after, the Cold War period. Actual and potential risks, associated with these sources for radioactive contamination and significant radiological consequences, mainly affect the conditions at local and regional levels in the European Arctic. This area consisting of the Barents Sea, the White Sea and the rim area including Svalbard, Novaya Zemlya, the Kola Peninsula and northernmost Norway, this is the most densely populated part of the Arctic with several cities (see Greaves in this volume). There live about 4 million inhabitants, of which almost one million are in, and just around, the Murmansk region of Russia.

There were several sources of radioactive pollution and nuclear accidents as potential risks, a few risky hotpots with relatively high probability of serious accident or environmental hazard, for societal security of Arctic residents and communities, as the following list of a state of radioactive sources and radiological hazards at the early-2000s well indicates (for more details see Heininen and Segerståhl 2002):

1. The Northern Fleet of Russia with about a hundred nuclear-powered submarines (under or out of operation) armed with nuclear weapons in several naval bases, as well as storage sites for nuclear warheads. Significant contamination has been every-now-and-then recognized and reported by media for example, at the Andreeva Bay naval base;
2. The Kola Nuclear Power Plant, Polyarnye Zori with two old reactors (VVER (PWR)-440/230 type) and two newer reactors (VVER-440/213 type);
3. Several nuclear ice-breakers in the headquarter harbor of RosAtom in Murmansk, and storage ships for radioactive waste;
4. Hundreds of storages for radioactive wastes and spent nuclear fuel;

5. Dumped radioactive waste and nuclear reactors in the Barents Sea, the Kara Sea and in the fjords of Novaya Zemlya, as well as sunken nuclear submarines with their reactors (e.g. Kursk submarine);
6. Building and repair shipyard in Severodvinsk (near-by Archangelsk) for the decommission of old nuclear submarines;
7. Nuclear weapons test ranges on Novaya Zemlya (117–132 tests done in 1950s–1990s), and underground nuclear explosions on the Kola Peninsula for civil purposes;
8. Testing of submarine-launched ballistic missiles (SLBMs) and torpedoes in the Barents Sea, as well as the rocket and missile launch site Plezesk in the Archangelsk region;
9. Transport of radioactive fuel, materials and wastes to the Russian nuclear installations in Siberia (Mayak in Chelyabinsk, Tomsk and Krasnoyarsk); and
10. Mining of radioactive lopartite near Lovozero (Fig. 7.1).

All in all, the nuclear problem of the Barents Sea area was complex and complicated by having the following special features and characteristics (they are not necessarily all exceptional, but the fact that the Barents Sea area included all of them can be taken as rather unique):

First, the Cold War legacy: One particular reason for the complexity of the nuclear problem in this remote region is the Cold War legacy. Due to the high important geo-strategic location of the Kola Peninsula with an open access to the Atlantic Ocean from its ice-free harbors, the military competition and arms race between the Soviet Union and the USA consisted of military structures, deployments, patterns of military action and also deterrence via nuclear weapons and facilities. They were a relevant part of the hidden agendas of the super-powers.

Second, a Russian problem: As a result of Stalin's modernization process the Soviet Union located its most toxic and hazardous industries in the Russian Arctic Zone. This together with the military-politically central location of the Kola Peninsula made the region highly strategic for the entire Soviet Union, as it is for the Russian Federation, as well as the Murmansk—St. Petersburg railway. Consequently, because this required more labour there was big migration of people from different part of the Soviet Union, and the region became one of the most populated parts of the entire Arctic. This is behind why environmental and socioeconomic impacts and risks of radioactive contamination had to taken into consideration, and nuclear safety became so important.

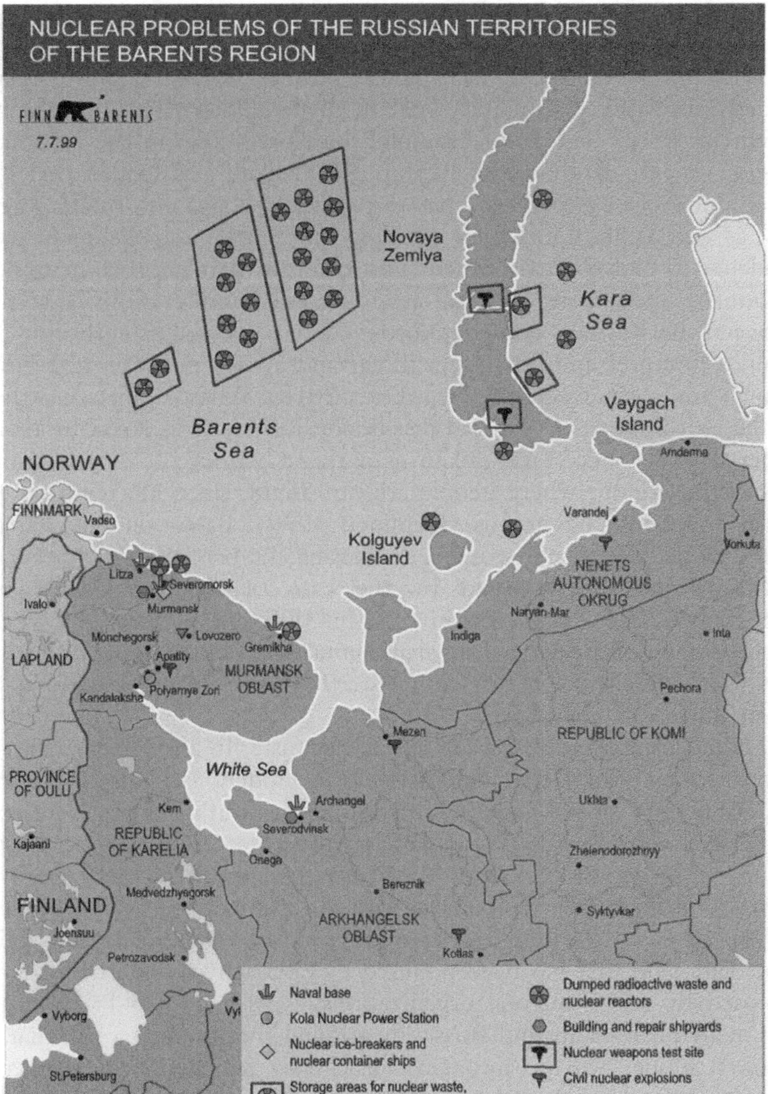

Fig. 7.1 Radioactive sources and hazards in the Barents Sea area at the early-2000s

Third, military and civilian processes: Following from the previous factor, the nuclear problem is closely connected to national security, which is underlined by the fact that the nuclear problem is mostly related to activities within the military system. Civil–military relations were mysterious to the West like for example, the closed cities in the Murmansk Region and the Arkhangelsk Region (Hönneland and Jörgensen 1999). There have been efforts to separate the military and non-military agendas, as well as hindrances, for example the military establishment had a tendency to "solve" environmental problems by denying their existence.

Fourth, there were, and still are, different interpretations of security between the Russians and the Nordic citizens, as well as authorities, on how to interpret and separate a threat and a risk, and how to consider human and societal security (e.g. Goes 2018): What is taken as a threat in the Nordic societies located just beside Russia, or as a risk by Nordic experts, is not necessarily taken as a threat within the Russian societies in the region, where people are not afraid, since life is interpreted to be inherently dangerous (Heininen 1999). Based on material collected from Russian newspapers, addressing the perceptions of security/environmental threat posed by the case of the Kursk submarine (Häyrynen 2003), however, gives a bit different picture: the nuclear problem, as well in general environmental impacts of military activities, was widely discussed in Russia in the 2000s due to the accident of the Kursk nuclear submarine.

Fifth, was international trade in nuclear waste: In 2000 the law not-to-import foreign radioactive materials was forbidden, and Russia started to import nuclear wastes from Germany, Japan, South Korea, Switzerland and Taiwan for burial on Russian territory. According to the Russian Ministry on Atomic Energy (Minatom) the money received from the storing of imported radioactive wastes—estimation of $20 billion—will be allocated to build proper store houses for nuclear wastes and to remediate land that has been polluted by radioactivity. The new law was opposed by environmental organizations, Russian Nobel Prize-winners and many protests around Russia by arguing that there is a risk that the received money will go into environmentally and politically dangerous projects (Popova et al., n.d.).

Sixth, multidimensional processes: There were many actors, both Russian and non-Russian, who were interested and active in nuclear safety issues, as well as other soft security issues (e.g. Pursiainen 2001; Heininen 1999). This meant on the one hand, multidimensional civilian and military processes with bureaucracy, and on the other hand, many steps in international cooperation to recognize the nuclear problem, as well as to include it as a part of international negotiations on nuclear safety. Efforts to solve the problem required a political will and ability from the Russian side to be open for international cooperation. As a result, technical and intellectual assistance and financial support from the international community (other European and Arctic countries) made possible to establish the Arctic Military Environmental Cooperation (AMEC) between Norway, Russia and the USA in 1996.

Seventh, there was considerable concern on the environment for large parts of North Europe and the Northern Atlantic due to trans-boundary pollution and a transfer from sources outside the region, such as radioactive contamination from atmospheric nuclear explosions at Novaya Zemlya, the radioactive deposition from the Chernobyl accident, and discharges from Sellafield, UK. For example, to the Icelanders and the Norwegians a nuclear accident in the main fishing areas of North Atlantic has been a real risk for decades, and thus a good reason for anxiety on the part of the Icelandic government already in the 1980s (Palsson 1988). Behind is the catastrophic influence of possible radioactive contamination to the Icelandic economy, dependent on fisheries, due to a serious nuclear accident in the Northern Atlantic.

Ironically, not only dumped radioactive wastes and nuclear submarine accidents impacted Arctic waters and partly the Northern Atlantic, but there were also other complicated toxics and sources of radioactivity. One of the major ones was technetium-99 from the Sellafied reprocessing plant in UK on the shore of the Irish Sea (e.g. *Newsweek* 2002). The leaks from Sellafield move from the Irish Sea, through the North Sea and along the Norwegian coast, to the Barents Sea and again down to the South along the eastern coast of Greenland—thus, radioactive pollution in the Barents Sea region was also coming from outside the region. Despite the bad reputation of Sellafield nuclear power reprocessing plant, which was known and protested against, and that there was international pressure from Ireland, Denmark and Norway toward the UK government to close the leaking Sellafield nuclear complex, there was political inability not openly discuss about the issue in international meetings, such as the Arctic Council Ministerial meetings, before the nuclear complex was closed in the early 2000s.

Finally, instead of the estimated slowness in the beginning, there has been good progress, since by January 2004 a total number of 117 nuclear submarines had been decommissioned and 60 of them have been dismantled, and more were waiting for dismantling (Ruzankin 2004). As a contribution of the AMEC cooperation there is now a proper and functioning technology to store radioactive materials and wastes. As the most impressive result, most parts of the Barents Sea area are almost clean in comparison to the end of the Cold War period, and the Barents Sea and Kara Sea largely exhibit very low concentrations of radioactive contamination. There in the Barents Sea area are still, however, risks and uncertainty locally due to the unsafe situation of stored radioactive wastes, which can be interpreted as a local threat or at least a concern for people and the environment.

All in all, in the end of the Cold War period the Barents Sea area had, and probably still has, the highest concentration of nuclear-armed and -powered submarines, nuclear infrastructure and nuclear wastes in the Russian Federation and the Arctic region, if not in the whole world. Behind is that nuclear safety is complex in the sense that issues are interconnected, frameworks are overlapping, scales are oscillating, and the more interaction there is, the more coordination is needed. In this context, the nuclear problem of the Barents Sea area is a multi functional and complicated issue. Furthermore, in that respect, the progress made with the Barents Sea area's nuclear problem is considered as an achievement, and first of all to be a result of international cooperation on environmental protection of and in the Arctic.

Process for a Transformation Towards Environmental and Societal Security in the Arctic

Thus, it was pollution, in particular radioactive contamination—'nuclear safety'—threatening peoples' everyday security in the 1980s, not climate change in the 2000s, which acted as the first trigger for environmental and societal security, as well as for significant geopolitical change in the entire Arctic region. These experiences were there when climate change came on the agenda and caused an 'awakening' of governments, and led the Arctic states to restructure their reality (Heininen 2018). Furthermore, this finally led peoples and communities to redefine security by identifying themselves as subjects of security, as Paci et al. (2004) states: "There is no civil security in a world where food safety, supply and quality, is uncertain".

In spite of international negotiations, treaties and declarations on nuclear safety aiming at reduction of nuclear risks coming from nuclear weapons and nuclear power plants, as well as citizens' activities, the concern of trans-boundary pollution and other environmental risks has not, however, always turned into legislation or common state-activities to stop pollution. Neither the concern nor agreements have been enough to stop arms race and proliferation of nuclear weapons and start a real nuclear disarmament, although the Cold War ended in the early 1990s (see Exner-Pirot in this volume). More important was the started process, which has been cumulative, and there are a few real outcomes, such as the following ones.

The first outcome was a growing concern on the state of the environment—the Arctic ecosystem—by non-state actors due to long-range air and water pollution, in particular radioactive wastes and nuclear accidents (Chernobyl, subs)—here nuclear safety is a part of environmental security. Following from this and pushed by an 'environmental awakening', the Nordic countries, in cooperation with other Arctic states, started functional cooperation for environmental protection originally based on one of the initiatives by President Gorbachev in his Murmansk Speech in 1987. This was applied by the Finnish government and its initiative led to the multilateral negotiation on environmental protection between the eight Arctic states, and finally to sign the AEPS in 1991.

The second one was that several scientific and expert reports on the state of the Arctic environment—among them the so-called Yablokov's "*White Paper*" (Yablokov et al. 1993), Bergman et al. *IIASA Report* (1996) on decommissioned nuclear subs as risky hotspots, AMAP reports on Arctic Environment (1997, 2002)—were published, and largely distributed and used.

Thirdly, since the process was cumulative soon after the signing of the AEPS the eight Arctic states, supported by Arctic Indigenous peoples and civil societies, established the Arctic Council, and the AMEC only two weeks after, in 1996. Furthermore, important international agreements on environmental protection, such as the Stockholm Convention on POPs, was signed having an epistemic community as major driver.

The fourth outcome was that, in addition to pollution, climate change became fast an issue on the political agenda of the Arctic states, after it was redefined as threatening peoples' everyday security, even state sovereignty, and scientifically proven to be accelerated by human activities. The rapid warming of the Arctic climate was interpreted, much due to 'environmental awakening' caused by pollution, as an emerging threat with risks of a race on natural resources, conflicts, as well as less room and resources for resilience.

Actually, there was a shift from environmental protection to climate change accelerated by the launch of the ACIA Report (2004): The focus of the Declarations of the AEPS and first Arctic Council Ministerial meetings (since 1991–2002), such as the Inari Declaration (in 2002) was "Action against pollutants", when that of the Reykjavik Declaration (in 2004) was "Climate change and the melting of ice". Interestingly, the focus of the Kiruna Declaration (2013) was "Improving economic and social conditions".

Followed from both pollution and climate change, the Arctic region, particularly the Arctic Ocean, became recognised as an environmental linchpin worldwide, first as a sink of long-range pollutants (AMAP 2002), and a bit later as a target area of rapid climate change (ACIA 2004; IPCC 2007; see also CARC et al. 2002). These developments pose new and more dangerous environmental and societal risks to the Arctic and its people(s) (Hoogensen Gjörv et al. 2013), but also have feedback effects on the social, technical, economic and political dynamics of global energy and natural resource systems.

Finally, it has been a cumulative process, since based on a new knowledge of pollution and climate change in the Arctic the states have called for robust international treaties concerning the environment, signed agreements on pollution prevention and worked towards better industrial management to minimise environmental risks. More importantly the Arctic states have firmly stated their commitment to "the well-being of the inhabitants…sustainable development in…the protection of the Arctic environment", and reaffirmed their "commitment to maintain peace, stability and constructive cooperation in the Arctic" (The Iqaluit Declaration 2015). Furthermore, the five Arctic Ocean coastal states, as well as the EU, Iceland, China, Korea and Japan, affirmed their commitment "to promote the conservation and sustainable use of living marine resources… in the Central Arctic Ocean" (The CAOF Agreement 2018).

By doing so, the Arctic states de facto adopted new, more comprehensive, discourses on security, such as environmental and human security, including food security. As well as, they caused and promoted a shift in their (traditional) security premises. However, so far the implementation of such mitigation efforts has been weak, since the Arctic states hesitate to take further steps. Due to this political 'inability' the states have not always been willing to adopt stricter environmental regulations for example, against the offshore petroleum industry, despite their political rhetoric for environmental protection to the contrary.

All in all, this has thoroughly changed Arctic geopolitics and governance, impacting several fields and factors, and upgrading environmental protection onto the core of the political agenda of Arctic cooperation (e.g. Heininen 2016, 2018). The process is much along to the risk society theory, and the discourse of comprehensive, or environmental security, as well as that first the environment, and then climate, were redefined as security factors and matters. Though not directly as 'a new discipline for disciplining', it caused a shift in security premises.

Changes in Arctic Security and Geopolitics—Special Features

It is a universal fact, almost a cliché, that what is guaranteed in the world, as well as in international politics/relations, is 'change'. Likewise, when facing wicked problems and trying to influence to your life and destiny, it is required to have a change. A change in problem definition can be interpreted as the most important factor in environmental politics (Haila 2001). And indeed, a change has also been a driving force of the 'politicization' of the environment.

Following from the transformation towards environmental and societal security it is possible to (re)conceptualize significant changes in Arctic geopolitics and Arctic security—here geopolitics and security are combined with each other—and analyze them. As outcomes of such exercise we see the following phases of the Arctic security nexus and changes in Arctic geopolitics (see also, Heininen 2016):

- The Cold War's security nexus (1950s–1980s): Meant first-of-all the hegemony competition between the Soviet Union and the USA (based on Technology models of Geopolitics). Dominated by traditional (national, military) security emphasizing the military, in particular the nuclear weapon systems (as advanced military technology). Led to militarization of the Arctic, as well as nuclear accidents and other environmental degradation by the military;
- Transition period's security nexus (1980s–1990s): Brought arms control and disarmament between the Soviet Union and the USA (a 'thaw'), as well as new kinds of security threats and new discourses (based on Risk Society theory and Constructivism). Dominated by the growing concern of nuclear accidents and radioactive wastes. Led to functional cooperation on environmental protection and nuclear safety (AMEC), and thus to apply 'Environmental security';

- Post-Cold War's security nexus (since 2000s): Brought geopolitical stability to balance military presence in globalization (based on Globalism), as well as accepted new security premises. Dominated by the "commitment to maintain peace, stability and constructive cooperation in the Arctic", state sovereignty/national security of the Arctic (littoral) states and human security (due to climate change). Led to geopolitical stability dis-spite/because of the heavy military (nuclear weapons) structures (for global deterrence) vis-à-vis impacts of climate change causing growing concerns vs. better access to exploitation ('Arctic paradox').

Correspondingly there have been significant changes in Arctic geopolitics:

- Cold War period's geopolitics: Much according to Classical geopolitics. Dominated by state hegemony and modernization, as well as ideological and political competition. Led to closed borders, minimum cooperation and a power game;
- Transition period's geopolitics: Dominated by concern for environmental degradation and transboundary cooperation on environmental protection. Led to establish intergovernmental (light) structures for cooperation in protection and research (e.g. AEPS, IASC), as well as growing self-consciousness and circumpolar cooperation by Arctic Indigenous peoples;
- Post-Cold War's geopolitics: Brought aspects of critical geopolitics, with its several factors, as well as more actors in. Dominated by climate change (as a wicked global environmental problem), hype of exploitation and global transport, and scientific vis-à-vis traditional knowledge (on climate, environment) as a new format of power. Led to reconstruction of the geopolitical reality of the Arctic states, as well as an emphasis on state sovereignty vis-à-vis international cooperation to maintain high geopolitical stability.

These two lists show how the environment, as well as comprehensive—environmental and human—security, was put onto the Arctic agenda, when, after the initiative of environmental and scientific cooperation by the President Gorbachev in 1987 (see Exner-Pirot, in this volume), the Arctic states first, affirmed their commitment to "*sustainable development in, … the protection of the Arctic environment*" (The Ottawa Declaration 1996); second, since 2011 have recognized "*the importance of maintaining peace, stability and constructive cooperation in the Arctic*" (The Nuuk Declaration 2011); and third, reaffirmed "*the commitment to maintain peace, stability and constructive cooperation in the Arctic*" (The Iqaluit Declaration 2015; The Fairbanks Declaration 2017).

Behind that was the environmental awakening and protests against pollution and global warming, and for environmental protection as universal trends and phenomena. This is based on a simple notion that pollution kills millions and causes cancer, and global warming threatens the everyday life of hundreds of millions. We are facing wicked global problems, such as climate change, which can destroy that material basis for human existence, and finally, the '*Anthropocene*' is at play.

Conclusions

As conclusions, the great success of international Arctic cooperation, based upon the functional cooperation on the protection of the Arctic environment, shows the importance of the environment. In particular, since this started at a time when the environment was considered a field of low politics, which only recently has been politicized. Furthermore, the environment is put into a focus, in rhetoric—though implementation is still partly on hold—by the Arctic states, Indigenous peoples and other Arctic actors. The environment and climate have become major factors of international Arctic cooperation. Consequently, in the post-Cold War era Arctic geopolitics and security are closely related to the environment, and the environment has become a special feature of Arctic security, as well as Arctic geopolitics.

The 'Arctic paradox' is not inevitable. Much depends on the Arctic states' criteria by which they make their decisions: To believe they can (re) construct their reality of post-Cold War Arctic geopolitics, since "anarchy is what states make of it" (Lamy et al. 2013). As mentioned earlier, already now more states are declaring climate change as a national security threat. There are the classical commitments vs. the reality of the situation: much depends on how security is (re)defined, who are interpreted as subjects of security, and if stability will be maintained! The relevant question is: Will the Arctic states, after recognizing climate change and environmental degradation as severe security factors and threats, be ready for an unorthodox process, i.e. a paradigm shift in approach, and a cultural shift to develop a mindset for resistance and resilience?

Not being dependent on a single solution, forgetting traditional ways (e.g. the jargon of sustainable development, legally-binding agreements), avoiding the "ecology as a new discipline for disciplining" moment, and following states' commitments, scientists' recommendations, INGOs' demands, and policy-makers' statements, an optimist with a certain amount of realism would interpret rapid climate change as the last warning, and an opportunity. Furthermore, if we are ready to consider these as global issues, ethical questions, and simply that there are learned lessons, the globalized Arctic—with experiences on environmental awakening implementing empowerment, nuclear safety causing a shift in security premises, and being a parameter for climate change—offers a common ground for political agency.

However, only if we go beyond 'political inability' and implement the original commitments, will the legitimacy among people be earned back. If we can adopt and apply a global perspective on how to look to the future and build on achievements, and utilize political stability as human capital, then security can be demystified, the current paradigm can be shifted towards "global environmental security" (The Inuit Declaration 2009), and Arctic actors can become forerunners in policy-shaping of the global politics of several uncertainties.

Note

1. This section of the chapter is based on the article of Lassi Heininen and Boris Segerståhl, "International Negotiations Aiming at a Reduction of Nuclear Risks in the Barents Sea Region". *Containing the Atom: International Negotiations on Nuclear Security and Safety*. Edited by Rudolf Avenhaus, Victor Kremenyuk, and Gunnar Sjöstedt. Lexington Books. International Institute for Applied Systems Analysis. New York, 2002, 243–270. See also, Häyrynen and Heininen (2002).

References

ACIA. 2004. *Impacts of a Warming Arctic: Arctic Climate Impact Assessment*. Cambridge: Cambridge University Press.

AMAP. 1997. *Arctic Pollution Issues: A State of the Arctic Environmental Report*. Oslo: Arctic Monitoring and Assessment Programme.

AMAP. 2002. *Arctic Pollution: Persistent Organic Pollutants, Heavy Metals, Radioactivity, Human Health, Changing Pathways*. Oslo: Arctic Monitoring and Assessment Program.

Beck, U. 1992. From Industrial Society to Risk Society: Questions of Survival, Social Structure and Ecological Enlightenment. In *Cultural Theory and Cultural Change*, ed. M. Featherestone. London: Sage.

Bergman, R., A. Baklanov, and B. Segerståhl. 1996. Overview of Nuclear Risks on the Kola Peninsula: Summary Report. *IIASA Radiation Safety of the Biosphere*. Laxenburg: International Institute for Applied Systems Analysis (IIASA).

The CAOF Agreement. 2018. The Agreement to Prevent Unregulated High Seas Fisheries in the Central Arctic Ocean. Signed on 3 October 2018 in Ilulissat, Greenland.

CARC, CMSS, and CPC. 2002. On Thinning Ice: Climate Change and New Ideas About Sovereignty and Security in the Canadian Arctic. Presentation Abstracts.

The Fairbanks Declaration. 2017. On the Occasion of the Tenth Ministerial Meeting of the Arctic Council, May 11, Fairbanks, Alaska. On-line available at http://www.arctic-council.org.

Finger, M. 2016. The Arctic, Laboratory of the Anthropocene. In *Future Security of the Global Arctic: State Policy, Economic Security and Climate*, ed. L. Heininen, 121–137. Basingstoke: Palgrave Macmillan.

Galtung, J. 1982. *Environment, Development and Military Activity: Towards Alternative Security Doctrines*. Oslo: Oslo-Bergen-Trondheim.

Goes, M. 2018. *Extracting Human Security from the Shtokman Gas Field: Security Assemblage in the Murmansk Region (2007–2012)*. PhD degree in Humanities and Social Sciences (Field Political Science) at UiT—Norwegian University of the Arctic (Mimeo).

Haila, Y. 2001. Johdanto: Mikä ympäristö? In *Ympäristöpolitiikka. Mikä ympäristö, kenen politiikka*, ed. Y. Haila and P. Jokinen, 9–20. Jyväskylä: Vastapaino.

Haila, Y., and L. Heininen. 1995. Ecology: A New Discipline for Disciplining? *Social Text* 42 (Spring): 153–171.

Häyrynen, N. 2003. Environmental Security: The Case of the Kursk. *Environmental Politics* 12 (3, Autumn): 65–82.

Häyrynen, N., and L. Heininen. 2002. Changes in Problem Definition—A Case Study of Nuclear Risks in the Barents Sea Region. In *Proceedings from the 5th International Conference on Environmental Radioactivity in the Arctic and Antarctic*, ed. P. Strand, T. Jölle, and Å. Sand, 113–115, June 16–20, St. Petersburg, Russia. Norwegian Radiation Protection Authority, Norway.

Heininen, L. 1999. The Nuclear Problem and Different Actors—The Problem onto a Political Agenda. In *Security in the European North—From 'Hard' to 'Soft'*, ed. L. Heininen and G. Lassinantti. Rovaniemi: Arctic Centre Reports 32.

Heininen, L. 2000. Sodan ja sotilaallisen toiminnan ympäristövaikutukset [Environmental Influences of War and the Military]. *Kosmopolis* 30 (4): 99–126.

Heininen, L. 2013. Politicization' of the Environment: Environmental Politics and Security in the Circumpolar North. In *The Fast-Changing Arctic: Rethinking*

Arctic Security for a Warmer World, ed. B.S. Zellen, 35–55. Calgary: University of Calgary Press.

Heininen, L. 2016. Security of the Global Arctic in Transformation—Changes in Problem Definition of Security. In *Future Security of the Global Arctic: State Policy, Economic Security and Climate*, ed. L. Heininen, 12–34. Basingstoke: Palgrave Macmillan.

Heininen, L. 2018. Arctic Geopolitics from Classical to Critical Approach—Importance of Immaterial Factors [PEEX Special Issue, ed. S. Chalov]. *Geography, Environment, Sustainability* 11 (1): 171–186. https://doi.org/10.24057/2071-9388-2018-11-1-1771-186.

Heininen, L., and M. Finger. 2017. The 'Global Arctic' as a New Geopolitical Context and Method. *Journal of Borderlands Studies* 33 (2): 199–202.

Heininen, L., and B. Segerståhl. 2002. International Negotiations Aiming at a Reduction of Nuclear Risks in the Barents Sea Region. In *Containing the Atom: International Negotiations on Nuclear Security and Safety*, ed. R. Avenhaus, V. Kremenyuk, and G. Sjöstedt, 243–270. New York: Lexington Books and International Institute for Applied Systems Analysis.

Hoogensen Gjörv, G., D.R. Bazely, M. Goloviznina, and A.J. Tanentzap. 2013. *Environmental and Human Security in the Arctic*. Tonbridge, Kent: Routledge and Eartscan from Routledge.

Hönneland, G., and A.-K. Jörgensen. 1999. *Integration vs. Autonomy: Civil–Military Relations on the Kola Peninsula*. Gateshead, Tyne and Wear: Ashgate.

The Inari Declaration. 2002. *The Third Ministerial Meeting of the Arctic Council*, October 10. Inari, Finland.

The Inuit Declaration. 2009. *A Circumpolar Inuit Declaration on Sovereignty in the Arctic*. Adopted by the Inuit Circumpolar Council.

IPCC. 2007. *Climate Change 2007: Synthesis Report. Summary for Policymakers. Intergovernmental Panel on Climate Change.*

The Iqaluit Declaration. 2015. Adopted at the Ministerial Meeting of the Arctic Council in May 2015 in Iqaluit, Canada. On-line available at http://www.arctic-council.org.

The Kiruna Declaration. 2013. Adopted at the Ministerial Meeting of the Arctic Council, May 15, Kiruna, Sweden. On-line available at http://www.arctic-council.org.

Lamy, S.L., J. Baylis, S. Smith, and P. Owens. 2013. *Introduction to Global Politics*, 2nd ed. New York, Oxford: Oxford University Press.

Lempinen, H. 2017. The Elusive Social? Remapping the Soci(et)al in the Arctic Energyscape. PhD dissertation at Faculty of Social Sciences, University of Lapland.

Nafeed, A. 2014. Nasa-Funded Study: Industrial Civilisation Headed for 'Irreversible Collapse'? *The Guardian*, International Edition, March 14.

Newcombe, H. 1986. Collective Security. Common Security and Alternative Security: A Conceptual Comparison. *Peace Research Reviews* 10 (3): 1–8 and 95–99.

Newsweek. 2002. Science&Technology, 44–45, February 11, "Waiting For Nukes".

The Nuuk Declaration. 2011. Adopted at the Ministerial Meeting of the Arctic Council in May 2011 in Nuuk, Greenland. On-line available at http://www.arcticcouncil.org.

The Ottawa Declaration. 1996. *Declaration on the Establishment of the Arctic Council*, September 19. Ottawa, Canada.

Paci, J., C. Dickson, S. Nikels, L. Chan, and C. Furgal. 2004. *Food Security of Northern Indigenous Peoples in a Time of Uncertainly*. Position Paper for the 3rd NRF Open Meeting, September 15–18, Yellowknife, NWT, Canada.

Palsson, T. 1988. Islannin ulko- ja turvallisuuspolitiikka. In *Paasikivi-Seuran monistesarja* nro 77. Helsinki: Paasikivi Society.

Popova, L., A. Yablokov, E. Kriusanov, and I. Gale. n.d. *Statement of Non-Governmental Environmental Organizations on the Plan to Export–Import Spent Nuclear Fuel* (Mimeo).

Pursiainen, C. 2001. *Soft Security Problems in Northwest Russia and Their Implications for the Outside World: A Framework for Analysis and Action* (Mimeo).

The Reykjavik Declaration. 2004. *The Fourth Ministerial Meeting of the Arctic Council*, November 24. Reykjavik.

The Rovaniemi Declaration. 1991. *Declaration on the Establishment of the Arctic Environmental Protection Strategy*. Rovaniemi.

Ruzankin, A. 2004. *Problems of Nuclear and Radiation Safety in the Murmansk Region*. Position Paper for the 3rd NRF Open Meeting.

United Nations. 1982. The Relationship Between Disarmament and Development. *United Nations, Study Series 5. A/36/597*. New York.

United Nations. 1983. Common Security. A Blueprint for Survival. *UN/A/CN.10/38*, Independent Commission on Disarmament and Security.

United Nations Development Programme, Human Development Report. 1994. Oxford: Oxford University Press.

Yablokov, A., V. Karasev, V. Rumyantsev, M. Kokeev, O. Petrov, V. Lystsov, A. Yemelyanenkov, and P. Rubtsov. 1993. *Radioactive Waste Disposal in Seas Adjacent to the Territory of the Russian Federation* (Mimeo).

CHAPTER 8

Conclusion: The Search for a New Security Paradigm Begins in the Arctic

Heather Exner-Pirot and Lassi Heininen

Abstract This chapter briefly summarizes the objectives of this book, and asks whether a new security paradigm—one that adequately accounts for the threats arising from environmental and climate changes—is on the horizon. We conclude that such a paradigm is not yet available. We remain entrenched in a state-led, international structure that is poorly adapted to address the security implications of climate change, as demonstrated in the Arctic region.

Keywords Arctic · Security · Climate change · Paradigm

This volume has sought to address the question of how we might begin to envision and move into a new security paradigm, given the confounding threats imposed by environmental and climate changes.

H. Exner-Pirot (✉)
Observatoire de la politique et la sécurité de l'Arctique at CIRRICQ, Quebec, Canada

L. Heininen
INAR at University of Helsinki, Helsinki, Finland
e-mail: lassi.heininen@ulapland.fi

L. Heininen and H. Exner-Pirot (eds.), *Climate Change and Arctic Security*, https://doi.org/10.1007/978-3-030-20230-9_8

Although there is growing agreement on how the premises of security have evolved, our responses to them—as states, individuals and communities—remain poorly matched for the threats imposed by new environmental insecurities.

The Arctic region offers potential insight into where security discourses might lead in the near future, having experienced already the impacts of and destabilization from environmental changes over the course of several decades. The impetus for regional cooperation in the Arctic can be traced to a desire to cooperatively address environmental threats, including nuclear waste, Persistent Organic Pollutants and other pollution, as manifested in the Arctic Environmental Protection Strategy in 1991 and the Arctic Council in 1996. However environmental protection has only moved from low to high politics in the past fifteen years. Despite the growing recognition of climate change, and even its securitization, state responses, not least in Arctic-specific fora, have been muted. Political efforts largely focus on understanding the changes and adapting to them, with little enthusiasm for large-scale mitigation efforts.

The chapters in this volume address different but complementary dimensions in the evolution of security discourses, particularly in the Arctic, to accommodate environmental threats. On the one hand, our conclusions are simple: anthropogenic environmental changes are being felt most intensely and most rapidly in the Arctic region, impacting not just ecosystems and northern peoples, but also infrastructure, economic models, and geopolitical dynamics. There is no definition of security for the Arctic—neither narrow nor broad, traditional nor non-traditional—that can avoid taking into account environmental aspects. This situation will only expand across the globe.

On the other hand, the situation is extremely complicated. Environmental security is not isolated from other security factors. China, for example, is both pushed and pulled into the Arctic due to the region's growing political and economic relevance as a result of climate change, however it has also put on display China's lack of leadership—or "climate responsibility"—on the issue. Oil and gas is both the cause of global and state interest in the Arctic region, and of its insecurity as a result of climate change. Cities are both a defense against the consequences of climate change, and the first victims of its impacts. Climate change has both heightened geopolitical tensions in the Arctic region, and caused Arctic states to collaborate more closely together. And while climate change is now pointed to as the source of all of these Arctic changes, pollution, in particular long-range ones, from nuclear waste and industrial activity fulfilled all these roles in the region and its security discourses in the 1980s and 90s, before climate change was even a political issue.

Having closely evaluated the evolution of security discourses in the Arctic region arising from climate change, as well as pollution, how confident are we that a new security paradigm—one that adequately accounts for the threats arising from environmental and climate changes—is on the horizon? Not very, at least not yet. Although there are earnest and productive efforts to shift security premises, as well as indicators of a shift, in the Arctic and elsewhere, the current state-led, international structure is poorly adapted for these new challenges. The Arctic provides insight into how environmental impacts are affecting security interests, and how the environment 'matters' when emphasizing the societal security of people and societies. But it does not provide, yet, a model for addressing them.

Index

L. Heininen and H. Exner-Pirot (eds.), *Climate Change and Arctic Security*, https://doi.org/10.1007/978-3-030-20230-9

The manufacturer's authorised representative in the EU is Springer Nature Customer Service Centre GmbH, Europaplatz 3, 69115 Heidelberg, Germany. If you have any concerns regarding our products, please contact ProductSafety@springernature.com

Printed and bound by CPI Group (UK) Ltd, Croydon, CR0 4YY
30/09/2026
02225260-0006